Food:

A Handbook

of Terminology, Purchasing,

and Preparation

ISBN-13: 978-0-8461-0005-8
ISBN-10: 0-8461-0005-3

Food:

A Handbook of Terminology, Purchasing, and Preparation

Nutrition, Health & Food Management Division
American Association of Family & Consumer Sciences

Eleventh Edition

American Association of Family & Consumer Sciences
1-800-424-8080   www.aafcs.org

Preface to the Eleventh Edition

The tenth edition served as a basis for the update of this edition of the Handbook. We used many pages intact, so the reader is indebted to previous authors and contributors, including Dr. Diane McComber, Dr. Rachel Schemmel, Dr. Nancy Lewis, Dr. Carolyn Bednar, CFCS, and Dr. Janice Harte. For the eleventh edition, Dr. Carolyn Bednar provided current information on the USDA dietary guidelines, labeling laws, and foodborne illnesses, as well as updates and additions to the sources of information provided in Chapter 14.

We welcome your comments and feedback on this book, which has been in print for 60 years! Please send them to AAFCS at staff@aafcs.org.

Dr. Carolyn Bednar, CFCS
AAFCS Nutrition, Health, and Food
Management Division Member
Professor and Chair, Department of Nutrition
& Food Sciences, Texas Woman's University

Table of Contents

General Food Information

CONSTRUCTION OF RECIPES

The writing of recipes carries with it a double responsibility. First, each recipe must be accurate and complete in essentials and written simply and clearly so that it cannot be misunderstood. At the same time it is important to present the recipe in a way that appeals to users while keeping within space limits.

A recipe is made up of two major parts—the list of ingredients and the method of preparation. Readily available ingredients, level measurements, and a simple procedure are ideal components of a recipe and should be used whenever possible. However, when the quality of a dish depends upon special ingredients or an unusual procedure, the recipe should also include this information.

The following are some well-tried rules for writing or editing recipes. Guidelines for metric usage also are presented. Since the metric system is used globally, this will facilitate conversion of recipe units for diverse groups.

INGREDIENTS

1. List all ingredients with measurements in the order used.

2. Do not abbreviate unless space is a problem.

3. Give ingredients in the fewest number of units of measure, as 1/4 cup instead of 4 tablespoons. Use standard measurements.

4. Use weights instead of measures when it is helpful, as for uncooked meat, poultry, fish, cheese and similar foods. Include weight or fluid measure for canned products.

5. Specify types of products needed, as *cake flour, all-purpose flour, dark corn syrup*.

6. For eggs, list *egg yolks or egg whites* or *eggs*, and include simple preparation, as *eggs, slightly beaten*. If special preparation is needed, as for meringue, explain in the method, detailing all necessary steps.

7. If possible, give the generic terms for products used. *(Many commercial firms will supply generic names for products upon request.)*

METHOD

1. Use short sentences and clear, simple directions that anyone can easily follow.

2. Give word pictures, like *chill until syrupy*, or *beat until foamy throughout*, or *mixture thickens as it cools*. These are helpful, particularly with unfamiliar mixtures.

3. Use methods for combining or cooking ingredients that represent best accepted procedures. For example, specify the best way to sift dry ingredients together, or to thicken a sauce, or to fold in beaten egg white—these can be expressed in the same terms whenever they occur in recipes. This practice results in a standardized consistency of phrasing, from recipe to recipe.

4. Give thought to the most efficient order of work to avoid using extra bowls, cups, measuring tools, extra beating, and so forth.

5. Specify standard sizes of baking pans or casseroles, as: *9-inch round layer pans, 1 1/2 inches deep* or shallow *1-quart casserole*.

6. Try to give both general and specific tests or temperatures. Then the recipe provides a double check on important stages. For example: *Cook to 238°F or until a small amount of syrup forms a soft ball in cold water.*

7. For the yield, give the number and size of servings to expect or the total measure, as: *Makes 4, 1-cup servings* or *1 quart*.

ACCURACY

1. Read over the edited recipe. Does it say exactly what is meant? Is it simple, clear, and complete, yet as brief as it can be?

2. Recheck the ingredients, amounts, and method against the original recipe data. Recheck the order of ingredients as listed against the order used. Check for any omissions in temperatures, times, yields, etc. Tell whether a canned ingredient should be drained or a pan covered during cooking.

3. With every word-processing of the recipe, read the new copy carefully against the former copy.

4. With every printing of the recipe, check first proofs against original correct copy, then proofread at each succeeding stage. Read proofs just for sense at least once.

RECIPE FORMS

The most used forms for presenting recipes are given here. Each has advantages.

Standard form. This familiar form gives all ingredients first, then the method. The listed ingredients show just what is needed to make the recipe.

In this form, when an ingredient is modified, the exact measurement should be given. For example: *2 cups sifted flour*, not 2 cups flour, sifted; *2 cups diced cooked carrots*, not 2 cups carrots, diced and cooked; *2 cups packed brown sugar*, not 2 cups brown sugar, packed. But, *1 cup heavy cream, whipped*, not 1 cup whipped cream.

The method follows in paragraphs or steps. This form is especially good for recipes using many ingredients. Following is an example of the *standard form:*

Soft Gingerbread

1 ²/₃ cups sifted cake flour
¹/₄   cup sugar
1   teaspoon double-acting baking powder
³/₄   teaspoon soda
¹/₂   teaspoon salt
¹/₂   teaspoon cinnamon
¹/₂   teaspoon ginger
¹/₄   teaspoon cloves
¹/₄   cup shortening (at room temperature)
¹/₂   cup molasses
¹/₂   cup water
1 egg, unbeaten

1. Preheat oven to 350°F (moderate).

2. Sift dry ingredients together into mixing bowl. Add shortening. Combine molasses and water and pour three-fourths of this mixture into mixing bowl.

3. Beat 2 minutes at low speed of electric mixer or 300 vigorous strokes by hand.

4. Add egg and remaining liquid, then beat 1 minute longer in mixer or 150 strokes by hand.

5. Pour batter into greased and lightly floured 9 × 9 × 2-inch pan.

6. Bake at 350°F 30 minutes, or until toothpick inserted in center comes out clean.

7. Serve hot with butter or lemon sauce.

Standard form in metric units. In converting from U.S. measurements to metric, the reader needs to realize that some countries use a capacity measurement (i.e., milliliters) for both wet and dry measurement while others use capacity for wet measurement and weight (grams) for dry measures. Additionally, tablespoons and cups in other countries can be different physical sizes from those used in the U.S., so recipes converted from metric might indicate those variances. Table 1.6 provides conversion from U.S. to metric and metric to U.S. for both capacity and weight.

The following recipe is a rather literal conversion, using acceptable replacement units with minimal rounding of measures to provide the metric measurements. The 250-milliliter (ml) measurer with 25-ml graduations is a standard and "measuring spoons" are available to measure 1, 2, 5, 15, and 25 milliliters. Dry measurers are 250, 125 and 50 ml. All recipes converted to metric units must be tested and adjustments made to provide the best product.

Soft Gingerbread

393 milliliters sifted cake flour
60 milliliters sugar
5 milliliters double-acting baking powder
4 milliliters soda
2 milliliters salt
2 milliliters cinnamon
2 milliliters ginger

1 milliliter cloves
60 milliliters shortening (at room temperature)
118 milliliters molasses
118 milliliters water
1 egg, unbeaten

1. Preheat oven to 180°C (moderate).

2. Sift dry ingredients together into mixing bowl. Add shortening. Combine molasses and water and pour three-fourths of this mixture into a mixing bowl.

3. Beat 2 minutes at low speed of electric mixer or 300 vigorous strokes by hand.

4. Add egg and remaining liquid, then beat 1 minute longer in mixer or 150 strokes by hand.

5. Pour batter into greased and lightly floured 23 × 23 × 5-centimeter pan.

6. Bake at 180°C 30 minutes, or until toothpick inserted in center comes out clean.

7. Serve hot with butter or lemon sauce.

Action form. This recipe style combines narrative action with listed ingredients. Although this style is easy to follow, it takes more space and is difficult to arrange economically or attractively on paper. The ingredients are described in the same way as in the standard form. Following is an example of the *action form*.

Soft Gingerbread

Preheat oven to 350°F (moderate).

Measure and sift together into mixing bowl:
1 ²/₃ cups sifted cake flour
¹/₄ cup sugar
1   teaspoon double-acting baking powder
³/₄ teaspoon soda
¹/₂ teaspoon salt
¹/₂ teaspoon cinnamon
¹/₂ teaspoon ginger
¹/₄ teaspoon cloves

Add ¹/₄ cup shortening (at room temperature)

Mix together:
¹/₂ cup molasses
¹/₂ cup water

Pour three-fourths of this molasses-water mixture into dry ingredients.

Beat 2 minutes at a low speed of electric mixer or 300 vigorous strokes by hand.

Add remainder of liquid and 1 unbeaten egg.

Beat 1 minute longer in mixer or 150 strokes by hand.

Pour batter into greased and lightly floured 9 × 9 × 2-inch pan.

Bake at 350°F 30 minutes, or until toothpick inserted in center comes out clean. Serve hot.

Descriptive form. In this recipe form, each ingredient is followed by the necessary modification. For example: *carrots, diced, cooked; cake flour, sifted; evaporated milk, whole or skim; brown sugar, packed; eggs, slightly beaten; process Cheddar cheese, if desired.* This enables the cook to see readily what ingredients are needed.

The amounts of the ingredients are given in a separate column. Each step in the procedure is a separate paragraph, which appears parallel to the ingredients involved. To save space, the procedure may be placed below the ingredients. This is the newest recipe form and is easy to follow. An example of the *descriptive form* (Hot Chicken Sandwich) illustrates this construction.

Hot Chicken Sandwich
6 servings

Toast, whole wheat, dry	6 slices	Preheat broiler
Chicken, cooked, sliced	¹/₃ to ¹/₂ pound, as desired	Place toast slices in a shallow baking pan. Cover toast with slices of chicken.
Cream of chicken soup, condensed, canned	10 ¹/₂-oz. can	Combine soup and milk in saucepan. Heat to simmering.
Evaporated milk	¹/₃ cup	Pour hot soup over sandwiches.
Tomato slices	6 large or 12 small	Arrange tomato slices on top of sandwiches
Bacon slices, cut in thirds, partly cooked	4	Top with bacon slices and olives
Olives, stuffed, sliced	6	Broil until sandwiches are hot and bacon is crisp.

Narrative form. This form includes the amounts of ingredients with the method. It is especially suited to the short recipe, the spoken recipe, or the recipe of few ingredients where the method is more complete. It can be expanded for detail or condensed for recipe ideas. Unless the recipe is short, this is the hardest form to follow, but it uses very little space. An example of the *narrative form* follows:

Baked Cod Fillets

Thaw one pound (500g) frozen cod fillets. Cut into serving portions. Arrange fillets in a well-greased baking dish. Brush with melted butter or margarine and sprinkle with salt and pepper. Pour 3/4 cup (175 ml) milk over fish. Combine 2 tablespoons (30 ml) melted butter or margarine with one cup (250 ml) soft bread crumbs and sprinkle over fish. Bake at 400°F (200°C) about 40-45 minutes or until fish flakes easily when tested with a fork. Makes 3-4 servings.

OVEN TEMPERATURE TERMINOLOGY

In giving oven temperatures most recipe writers state the degrees first and follow with the descriptive term, thus, 400°F or 200°C (hot) oven. The following listing shows the commonly accepted descriptive terms for each temperature range in degrees Fahrenheit and Celsius:

250°F to 275°F .... 120°C to 140°C .... Lukewarm oven

300°F to 325°F .... 150°C to 160°C .... Warm oven

350°F to 375°F .... 180°C to 190°C .... Moderate oven

400°F to 425°F .... 200°C to 220°C .... Hot oven

450°F to 475°F .... 230°C to 240°C .... Very hot oven

500°F to 525°F .... 260°C to 270°C .... Extremely hot oven

MODIFICATION OF RECIPES TO MEET DIETARY GUIDELINES

Since the link between diet and health is so important, the U.S. Department of Agriculture (USDA) and U.S. Department of Health and Human Services (HHS) periodically update and release **Dietary Guidelines for Americans.**

Some key messages of the 2005 Dietary Guidelines include:

- Consume nutrient-dense foods from all food groups.
- Limit intake of saturated and trans fats, cholesterol, added sugars, salt, and alcohol.
- Be physically active every day.
- Choose a variety of fruits and vegetables, variety of grains, including whole grains, and fat-free or low-fat milk and dairy products.

USDA has also produced a revised Food Pyramid called MyPyramid that translates the Dietary Guidelines into specific dietary recommendations for individual consumers. MyPyramid focuses on the total amount of food an individual should consume from each food group daily (ounces or cups) instead of recommending a specific number of servings. For detailed information, visit both the Dietary Guidelines (www.health.gov/dietaryguidelines) and MyPyramid (www.mypyramid.gov).

Most people can benefit by modifying what foods they choose to eat. However, another way to modify diet is to make adjustments in the types and amounts of ingredients in recipes to produce more nutrient dense food products (i.e., lower in fat or salt, higher in fiber).

WHEN TO MODIFY RECIPES

Not all recipes need to be modified. Consider the following questions.

1. **Is the recipe already low in fat, cholesterol, sugar or salt?**

   If so, only minor or no changes may be needed. For example, if a recipe calls for an egg, and the dish serves eight people, the amount of cholesterol per serving already is low.

2. **How often is the food eaten?**

   It is not as important to modify a recipe for a dish eaten once or twice a year as it is for food eaten frequently. For example, it is more important to cut the fat in a weekly tuna fish salad sandwich than it is to cut the fat in a birthday cake.

MyPyramid Food Guidance System

MyPyramid
STEPS TO A HEALTHIER YOU
MyPyramid.gov

GRAINS	VEGETABLES	FRUITS	MILK	MEAT & BEANS
GRAINS Make half your grains whole	**VEGETABLES** Vary your veggies	**FRUITS** Focus on fruits	**MILK** Get your calcium-rich foods	**MEAT & BEANS** Go lean with protein
Eat at least 3 oz. of whole-grain cereals, breads, crackers, rice, or pasta every day 1 oz. is about 1 slice of bread, about 1 cup of breakfast cereal, or ½ cup of cooked rice, cereal, or pasta	Eat more dark-green veggies like broccoli, spinach, and other dark leafy greens Eat more orange vegetables like carrots and sweetpotatoes Eat more dry beans and peas like pinto beans, kidney beans, and lentils	Eat a variety of fruit Choose fresh, frozen, canned, or dried fruit Go easy on fruit juices	Go low-fat or fat-free when you choose milk, yogurt, and other milk products If you don't or can't consume milk, choose lactose-free products or other calcium sources such as fortified foods and beverages	Choose low-fat or lean meats and poultry Bake it, broil it, or grill it Vary your protein routine — choose more fish, beans, peas, nuts, and seeds
For a 2,000-calorie diet, you need the amounts below from each food group. To find the amounts that are right for you, go to MyPyramid.gov.				
Eat 6 oz. every day	Eat 2½ cups every day	Eat 2 cups every day	Get 3 cups every day; for kids aged 2 to 8, it's 2	Eat 5½ oz. every day

Source: U.S. Department of Agriculture.

3. How much of the food is eaten?

Sometimes the best way to modify the intake of a certain food is to eat less of it. Decreasing the quantity eaten may be more satisfying than altering the recipe to contain less fat or kilocalories.

HOW TO MODIFY RECIPES

1. Identify the dietary goal and decide which nutrients to modify. To reduce kilocalories (joules), identify ingredients that contribute the most calories (joules), usually fat and/or sugar. To reduce sodium or to increase fiber, identify the ingredients that contain these components.

Kilocalories: Fat is the most concentrated source of kilocalories (joules). Each gram of fat supplies nine kilocalories (38 kJ), compared to seven kilocalories (17 kJ) per gram of alcohol. Reducing the amount of fat in a recipe is the most effective way to cut kilocalories (kJ).

Fat. Fatty acids are basic components of fat. They can be classified as saturated, trans, monounsaturated, or polyunsaturated. All fats in foods contain mixtures of these fatty acids. Consumption of high amounts of saturated fats and trans fats raises the level of serum cholesterol and increases risk of cardiovascular disease for individuals.

Saturated fats are found in the largest proportions in fats of animal origin such as whole milk, cream, cheese, butter, lard, meat and poultry. They also occur in large amounts in coconut oil and palm kernel oil.

Monounsaturated fats are found in large amounts in canola oil, olive oil and peanut oil, as well as in many margarines and semi-solid vegetable shortenings.

Polyunsaturated fats are found in the largest proportions in fats of plant origin such as liquid vegetable oils (safflower, sunflower, corn, cottonseed and soybean) and margarines and salad dressings made from them.

Trans fatty acids are formed when oils are partially hydrogenated to produce more solid fats such as margarine or hydrogenated shortening. This process increases shelf life, flavor stability, and melting point for these fats. Trans fats are found mainly in processed foods made with partially hydrogenated shortening or margarine including crackers, cookies, cakes, pies, snack foods, and other baked foods. Foods that are fried in partially hydrogenated shortening such as French fried potatoes are also high in trans fat content.

Cholesterol is a fat-like substance found only in food of animal origin. Sources include egg yolks, meats (particularly organ meats), butter, cream, and cheese.

Sugar comes in many forms, for example, white sugar, brown sugar, raw sugar, honey, corn syrup, molasses and maple syrup.

Sodium is present in salt (salt is 40 percent sodium), baking soda, baking powder, monosodium glutamate (MSG), soy sauce, bouillon, pickles, olives, cured meats, many canned vegetables and conve-

nience foods, most cheeses, sauces, and soups.

Fiber is present in whole grain breads and cereals; dry beans and peas; nuts and seeds; and fruits and vegetables, especially those with edible skins or seeds.

2. **Determine how to change the ingredient(s) to achieve the dietary goal.** Ingredient(s) can be:

   a. eliminated completely,
   b. reduced in amount, or
   c. substituted for a more nutritionally acceptable ingredient.

   To choose the best approach, it is helpful to have a general idea of the function of the ingredients, and what will happen if they are modified. Table 1.1 (page 9) lists the function and nutritional contribution of several ingredients.

TIPS FOR HEALTHY MODIFICATIONS

The following are a few ways to modify recipes. These suggestions can be applied to most food; exceptions include those in which specific proportions of ingredients are essential to prevent spoilage or to create a certain product consistency (such as salt in cured meats or pickles, or sugar in jams and jellies).

1. **To Decrease Total Fat and Kilocalories:**

   - Reduce fat by one-fourth to one-third in baked products. For example, if a recipe calls for 1 cup (250 ml) hydrogenated shortening, try ²/₃ cup (165 ml). This works satisfactorily in many quick breads, muffins and cookies.

   - Replace one-third to one-half of fat in baked products by substituting bananas, applesauce, or prune puree.

   - Sauté or stir-fry vegetables with little fat or simmer in water, wine, or broth.

   - Steam or microwave vegetables.

   - To thicken sauces and gravies without lumping, eliminate fat and instead mix cornstarch or flour with double the measured amount of cold liquid. Stir this mixture slowly into the hot liquid to be thickened and bring it to a boil, stirring constantly.

Add herbs, spices, or flavorings.

- Chill soups, gravies, and stews, and skim off hardened fat before reheating to serve.
- Select lean cuts of meat and trim off visible fat. Remove skin from poultry before cooking.
- Bake, broil, grill, poach, or microwave meat, poultry, or fish instead of frying in fat.
- Decrease the proportion of oil in homemade salad dressings. Try one-third oil to two-thirds vinegar. Low-fat cottage cheese, yogurt and buttermilk seasoned with herbs and spices also make low-fat dressings.
- Use reduced-calorie or nonfat sour cream, mayonnaise, plain low-fat or nonfat yogurt, buttermilk, or blended cottage cheese instead of regular sour cream or mayonnaise for sauces, dips and salad dressings. If a sauce made with yogurt is to be heated, blend one tablespoon of cornstarch with one cup of yogurt to prevent separation.
- Use nonfat or low-fat milk instead of whole milk. For extra richness, add undiluted evaporated skim milk.
- Choose low-fat cheeses such as feta, Neufchatel, and mozzarella instead of high-fat ones such as Swiss or Cheddar. Also use less cheese.

2. To Decrease Saturated Fat and Cholesterol

- Use two egg whites or an egg substitute product instead of one whole egg. In some recipes, simply decrease the total number of eggs. This is especially true if the fat is also decreased in the recipe.
- Use olive oil on breads/toast instead of margarine or butter. Otherwise, look for margarines in which liquid vegetable oil is the first ingredient listed on the label.
- Use vegetable oils instead of solid fats. To substitute liquid oil for solid fats, use about one-fourth less than the recipe calls for. For example, if a recipe calls for 1/4 cup or 4 tablespoons (60 ml) of solid fat, use 3 tablespoons (45 ml) of oil. For cakes or piecrusts, use a recipe that specifically calls for oil, because liquid fats require special mixing procedures.

3. To Decrease Sugar

- Reduce sugar by one-quarter to one-third in some baked goods and deserts. Add extra spice or flavoring to enhance impression of sweetness. This works best with quick breads, cookies, pie fillings, custard, puddings, and fruit crisps.
- Decrease or eliminate sugar when canning or freezing fruits. Buy unsweetened frozen fruit or fruit canned in its own juice or water.
- In cookies, bar and drop, replace one-quarter of sugar called for with an equal proportion of nonfat dry milk; this reduces kilocalories and increases calcium, protein, and riboflavin in the recipe.
- Choose fruit juices, club soda or skim milk over soft drinks and punches. Make fruit juice coolers with equal parts fruit juice and club soda or seltzer.
- Nonsugar sweeteners now on the market include aspartame, acesulfame potassium, saccharin, and sucralose. All of these sugar substitutes work well in some applications, but acesulfame potassium and saccharin are not recommended for use in baked products. Both aspartame and sucralose are now marketed as sugar blends for baking that produce acceptable results with a small reduction in calories. The beginning cook is advised to use recipes especially developed for these products.

4. To Decrease Sodium

- Salt may be omitted or reduced in most recipes. Do not reduce the salt in cured meats or pickled or brined vegetables, because it acts as a preservative. A small amount is also useful in yeast breads to help control the rising action of the yeast.
- Start with a gradual reduction. For example, if a recipe calls for 1 teaspoon (5 ml) of salt, try 1/2 teaspoon (2 ml). If the amount of salt is reduced gradually, an individual will soon adjust taste preferences to foods containing less salt.
- Choose fresh or low-sodium versions of products. For example, choose low-sodium soups and broths, soy sauce, canned vegetables, and tomato products.
- Rely on herbs and spices rather than salt for flavor.
- Use garlic or onion powder instead of garlic or

onion salt.

- Reduce the salt used in cooking pasta, noodles, rice or hot cereals.

- Use water from the cold water tap for cooking. Water softeners contain sodium and are frequently used to prepare hot water more suitable for cleaning purposes.

- Dilute broth or bouillon with water when using as the cooking liquid.

- Read labels. Any ingredient that includes "sodium" as part of its name contains sodium; mgs of sodium are listed as part of the labeling information.

5. To Increase Fiber

- Choose whole grain instead of highly refined products (for example, whole-wheat flour and bread, bulgur, brown rice, oatmeal, whole-corn meal, and barley).

- Whole-wheat flour usually can be substituted for up to one-half of the all-purpose refined flour in a recipe. For example, if a recipe calls for 2 cups (500 ml) of all-purpose flour, try 1 cup (250 ml) of all-purpose and 1 cup (250 ml) of whole-wheat flour.

- Add extra fruits and vegetables to recipes.

- Add fruits to muffins, pancakes, salads, and deserts, and add vegetables to quiche fillings, casseroles, and salads.

PUTTING IT INTO PRACTICE

To begin, look at the ingredients in the recipe and review their functions (Table 1.1). Then look at the general guidelines for modifying ingredients. Use the following steps when modifying a baked product.

1. **Compare the recipe with the basic recipe proportions given in this book**. To do this, express ingredients in terms of 1 cup of flour (250 ml). For example, if the recipe calls for 2 cups of flour (500 ml) and 1/2 cup sugar (125 ml), it calls for 1/4 cup sugar (60 ml) for every 1 cup flour (250 ml).

2. **Decide which ingredients to reduce**. Compare the amounts of fat, eggs, sugar and salt per 1 cup (250 ml) flour with amounts given in a basic recipe pro-

portion table. Those on the high end of the range are the ones to reduce. Reduce fat and/or sugar to at least the midpoint of the range. Reduce eggs, if needed, to within range. Reduce sodium if desired.

3. **Decide which ingredients to increase**. Fat and sugar provide moistness and richness to recipes. When they are reduced, liquid needs to be added, usually in the form of water, milk, fruit juice, applesauce, or fruit or vegetable pulp. Fruit and vegetable pulp also add fiber; milk adds calcium. Add back at least half as much liquid as the volume of reduced sugar and fat reduced.

For example, if the sugar is reduced in a recipe by 2 tablespoons (30 ml) and fat by 4 tablespoons (60 ml), add 3 tablespoons (45 ml) milk or fruit juice. Because honey and molasses have greater moisture retaining properties than sugar, replacing a portion of the sugar that remains in the recipe with honey or molasses may improve softness and moistness in the end product.

4. **Consider adding nonfat dry milk to increase calcium content**. In many baked products, nonfat dry milk can replace up to one-fourth of the sugar in the recipe with good results. Two to 4 tablespoons (30 to 60 ml) of nonfat dry milk powder per cup (250 ml) of flour also may be added to cookies and cakes as an additive. Look at the recipe and add nonfat dry milk powder equal to the reduction in sugar.

5. **Try recipe with modifications made**. Use the toothpick or touch test, not visual appearance, to evaluate doneness. Because sugar slows protein coagulation and slows starch gelatinization, products made with reduced sugar will bake more quickly and be less brown when done.

6. **Evaluate the quality and taste of the product**. If the recipe still seems richer than desired, make further reductions in fat and sugar and increase liquid. If the product seems dry, be careful to not over bake. If still dry and not over baked, try adding more liquid (water, fruit juice, applesauce, fruit, or vegetable pulp), or substitute a portion of the sugar in the recipe with honey or molasses.

TABLE 1.1

Function and Nutritional Contribution of Baking Ingredients

Ingredient	Function	Kilocalories (Kilojoules)	Nutritional Value
Flour	Contributes structure and rigidity. Gives stretch or elasticity. Provides thickening.	460/cup (1925)	Provides complex carbohydrates, some protein, source of B vitamins, and iron if enriched or whole grain.
Liquid	Dissolves dry ingredients (flour, sugar, salt). Moistens flour to develop structure. Dissolves leavening agents, e.g., baking powder, baking soda.	Water = 0/cup Milk = 80-160/cup (335-670)	Milk is a good source of calcium, riboflavin, protein; also provides some vitamin A and thiamin. Whole milk contains some saturated fat.
Salt	Helps improve taste. Prevents large air cells in yeast products by controlling rising.	0 units	Provides 2,000 mg. sodium/tsp. (5 ml) and 3000 mg. chloride.
Leavening Agents	Important for structure and lightness.	0 units	Baking powder and soda contain sodium; yeast does not.
Sugar	Adds sweetness. Increases volume. Helps tenderize product. Aids in browning.	Sugar = 770/cup (3222) Honey = 1020/cup (4269) Molasses = 700-825/ cup (2930-3453)	Provides energy. Molasses provides some calcium and iron.
Fat	Helps tenderize. Adds flavor to product. Increases moistness.	Lard, shortening, oil = 135/Tbsp (520), 2160/cup (8032) Butter/Margarine = 110/Tbsp 1760/cup (460-7366)	Provides fat calories and fat soluble vitamins. Butter, lard, vegetable shortening and palm and coconut oil are higher in saturated than polyunsaturated fat. Soft margarine and other oils are higher in polyunsaturated fat. Butter, lard, and other animal fats contain cholesterol. Butter and margarine provide some sodium and vitamin A. Vegetable oils provide vitamin E.
Eggs	Contribute to elasticity and structure. Beaten whites incorporate air into batters. Yolks add color and emulsify fats.	80/egg (335)	Provide protein and small amounts of several nutrients. Egg yolk rich in fat, saturated fat, and cholesterol.

(continued on next page)

TABLE 1.1 (CONTINUED)

Function and Nutritional Contribution of Baking Ingredients

Ingredient	Function	Kilocalories (Kilojoules)	Nutritional Value
Nuts	Flavor and variety.	750-850/cup (3139-3557)	Provide some protein, B vitamins, iron, and fiber. Generally rich in fat and polyunsaturated fat.
Fruits and Vegetables	Flavor and variety.	Fruits = 100-200/cup (418-837) Vegetables— 20-150/cup (84-628)	Generally good source of B vitamins and vitamins A and/or C. Good source of fiber, especially when skins and tiny seeds are consumed.

7. **Try reformulated recipe and repeat steps 5 and 6 until pleased with results**. Remember to record the modifications that work. Recipes that meet the Dietary Guidelines are worth repeating. The Carrot Bread (on the following page) is an example of an original recipe with modifications.

HIGH-ALTITUDE FOOD PREPARATION

At altitudes above 3000 feet (914 m) food preparation by traditional, microwave, or campfire methods may require changes in time, temperature, recipe proportions, microwave power levels and/or pan size. The reason is reduced atmospheric pressure at higher elevations; the pressure decreases about 1/2 pound per 1000 feet or 2.2 Newtons per 305 meters increase in elevation. Pressure per square inch (0.645 square meter) at sea level is 14.7 pounds (6.7 kg), at 5000 feet (1524 m) is 12.3 pounds (5.6 kg), and at 10,000 feet (3048 m) is 10.2 pounds (4.6 kg). This decreased pressure affects food preparation in several ways:

- Water and liquids boil at lower temperatures and evaporate faster.
- Leavening gases in breads and cakes expand more.
- Liquids foam more quickly, and cakes and breads rise more during baking; liquid or cakes may flow over the sides of the pan.
- Microwaved foods thaw and reheat at a slightly slower rate.

- Foods microwaved at 5000 feet (1524 m) versus sea level are often more moist, but they cool and dry out rapidly. Store baked goods covered.
- Foods cook more slowly in boiling water as altitude increases.
- Approximate boiling temperatures of water at various elevations:

Sea level	212.0°F/100.0°C
2000 feet/610 meters	208.2°F/ 97.8°C
5000 feet/1524 meters	202.5°F/ 94.5°C
7500 feet/2286 meters	197.8°F/ 91.8°C
10,000 feet/3048 meters	193.0°F/ 89.0°C

Each 1000 feet (304.8 m) increase in elevation decreases the boiling point temperature 1.9°F (1.1°C).

BOILED, COOKED, AND FRIED FOODS GUIDELINES

Since water boils at lower temperatures at higher altitudes, food products require a longer cooking time to reach the desired end point.

Beverages/soups: Increase the amount of coffee/tea and brew time for similar flavors. Brew in pot rather than cup to release flavors and maintain heat. Above 8000 feet (2438 m), brewed beverages are not very hot. Homemade soups need more time and water. Milk-based beverages boil over easily.

Vegetables: Increase boiling time 4−11 percent at 5000 feet (1524 m), 20-25 percent at 7200 feet (2194 m), and

Carrot Bread

Original Recipe	Kcal/(kJ)	Modified Recipe	Kcal/(kJ)	Improvement(s) Made
Sugar, 3/4 cup (175 ml)	578 (2419)	1/4 cup sugar (60 ml)	192 (804)	Decreased sugar
		1/2 cup orange juice (125 ml)	60 (251)	Increased vitamin C, folic acid, and moisture
Soybean oil, 1 cup (250 ml)	1920 (8035)	1/3 cup corn oil (80 ml)	636 (2662)	Decreased fat, kilocalories
Eggs, 2	160 (670)	One egg	80 (335)	Decreased cholesterol
Vanilla, 1 1/2 tsp. (7 ml)	0	Same	0	_____
All-purpose flour, 2 cups (500 ml)	910 (3808)	1 cup all-purpose flour (250 ml)	862 (3607)	Increased fiber and trace minerals
		1 cup whole wheat flour (250 ml)		
Baking Powder, 1 tsp. (5 ml)	0	Same	0	_____
Baking Soda, 1/2 tsp. (2 ml)	0	Same	0	_____
Cinnamon, 1 Tbsp (15 ml)	0	Same	0	_____
Salt, 1/2 tsp. (2 ml)	0	Salt, 1/4 tsp. (1 ml)	0	Decreased sodium
Carrots, grated 1 1/4 cups (300 ml)	50 (209)	Carrots, grated 1 1/2 cups (375 ml)	60 (251)	Increased vitamin A, beta-carotene, and fiber
TOTALS: Recipe	3618 (15141)		1890 (7910)	
1/2-inch slice (1.25 cm)	201 (841)		105 (439)	

Directions: Beat together sugar, oil, eggs, orange juice, and vanilla. Stir dry ingredients together and add to sugar and oil mixture. Blend in grated carrots. Bake in oiled 9 × 5-inch (23 × 13 cm) loaf pan at 350°F (180°C) for 40 to 50 minutes.

55-66 percent at 10,000 feet (3048 m) and add more water. Foil-wrapped potatoes with roast at 325°F (160°C) need 2-2½ hours at 5000 (1524 m) and 7500 feet (2286 m), and 3 hours at 10,000 feet (3048 m). Prick potatoes to prevent explosions during baking. A general rule for microwaving fresh whole vegetables is to heat on high for 7 minutes/pound (454g) for cut vegetables, then add 1-2 tablespoons (15-30 ml) water and microwave ½-1 minute more. For frozen vegetables, add 2-4 tablespoons (30-60 ml) water and 1-3 minutes more time. To prepare dried beans, soak overnight, and then cook, using 2-3 times more water and greatly increased time; a pressure saucepan is recommended for cooking soaked, dried beans at altitudes above 7500 feet (2286 m).

Meat/poultry: For moist heat cooking at 5000 feet (1524 m) 25-50 percent more time is needed for braising or simmering meat and 5 to 35 minutes more time for microwaving. Watch the liquid levels, and use tight fitting lids. Microwave for the first period on High for 8-10 minutes until boiling, then cook at lower power until fork tender. For dry heat roasting in a hot oven, use the same or slightly longer cooking times; check for doneness with a thermometer. Large or dense pieces may need 10-25 percent more time. A three-pound (1.36 kg) chicken needs 10-15 minutes more roasting time or 3-5 minutes more when microwaving. Pan-fry at slightly higher settings; increase electric skillet/wok temperature by 25°F (14°C). For grilling, a partial precook and frequent basting are helpful to prevent drying of meat. NOTE: Charcoal briquettes need 45 minutes to become 70 percent ashed-over at 5000 feet (1524 m) versus 20-25 minutes at sea level. In the microwave, one pound (454 g) ground meat needs 1-3 minutes more time.

Fruits and steamed puddings: More time for all cooking methods may be needed. Above 7500 feet (2286 m), precooking pie fruit is desirable.

Egg cooking:
Soft cooked

5000 feet/1524 meters	5-6 minutes
7500 feet/2286 meters	7 minutes
10,000 feet/3048 meters	9 minutes

Hard cooked

5000 feet/1524 meters	20-25 minutes
7500 feet/2286 meters	23-27 minutes
10,000 feet/3048 meters	26 minutes

Slight time increases are needed for poached eggs.

Sugar cooking: Candies, syrups, frostings, and jellies need a decrease in the final cooked-stage temperature. The temperature should be decreased by the difference in the boiling water temperature at the preparation altitude and that of sea level. Example: A thermometer in boiling water reads 202°F (94.5°C) or ten degrees below 212°F (5.5 below 100°C). For soft-ball candy, subtract 10 degrees from 238°F (114°C) and cook only to 228°F (5.5 from 114.5°C or 109°C). The cold water candy and jelly sheeting tests are the same as at sea level. Microwaved candies foam up more, so use a larger dish or make half of the recipe.

Starch Cooking: At altitudes of 5000 feet (1524 m) or higher, problems may be encountered in the use of cornstarch for thickening. Direct heat, a heavy saucepan (not a double boiler) and constant stirring to prevent scorching are needed for the starch to reach maximum gelatinization. Above 7500 feet (2286 m), replace 1 tablespoon (15 ml) of cornstarch with 2 tablespoons (30 ml) flour in pudding/filling. When microwaving, make sure the whole mixture is boiling and thickened; add time if necessary.

Rice/pasta: Add 1 teaspoon (5 ml) oil per cup of rice/pasta to prevent sticking and boiling over. For rice at 5000 feet (1524 m) add an additional 4-8 tablespoons (60-120 ml) water per 1 cup (236.6 ml) of water in the recipe and increase time by 25-30 percent. When microwaving, cook on high 5-8 minutes until boiling and continue at 50-percent power, increase time by 30-40 percent. For pasta, keep water at an adequate boil, taste test and generally add more cooking time. When microwaving cereals, use a large bowl. In casseroles with rice, pasta, dried beans and firm vegetables, precook these items. For milk-based casseroles, use a larger dish to prevent boiling over.

Deep-fat frying: The lower boiling point of water in foods requires lowering the temperature of the fat to prevent overbrowned but raw foods. A rough guide is to lower the frying temperature about 2-3°F for each 1000 feet (1 to 1.5°C per 304.8 m) increase in elevation. Example:

At 8000 feet (2438 m), doughnuts need frying fat temperatures at 359°F to 351°F for 24 to 16°F (183 to 177°C or 13 to 7°C) lower. Cake-like doughnuts may need less baking powder, sugar and/or shortening.

Canning, pressure canning and refreezing: See Section 13, Food Preservation. Pressure saucepans designed only for pressure-cooking should not be used for canning.
Pressure saucepans: Pans which are designed ONLY for pressure cooking and NOT canning require an increase in pressure to achieve the required processing temperature.

Steam Pressures at Various Altitudes

Altitude		Pressure Required for 240°F, 116°C		Pressure Required for 250°F, 121°C	
Feet	Meters	lb.	kPa	lb.	kPa
Sea Level	0	10	68.9	15	103.4
1000	305	10.5	72.4	15.5	106.8
2000	610	11	75.8	16	110.3
3000	914	11.5	79.3	16.5	113.8
4000	1219	12	82.7	17	117.2
5000	1524	12.5	86.2	17.5	120.7
7000	2134	13.5	93.1	18.5	127.6
10000	3048	15	103.4	20	137.9

LEAVENED AND BAKED FOODS GUIDELINES

Since the atmospheric pressure is less at higher elevations, leavened baked products can rise excessively, stretching the cell structure, producing a coarse, irregular texture, overflowing the pan sides, and even falling in the center with a sway or depression and low volume. Quick breads, if low in sugar, are firm enough to withstand the excess rising and need only to have leavening reduced, liquid added and temperature and maybe pan size increased. For cookies, see "Hints" on page 14. Generally microwaved cakes require fewer adjustments than do cakes baked in an oven. Most cakes, cake-like products and breads are affected slightly at 2000 feet (610 m) and significantly above 3500 feet (1067 m). Adjustments in the recipe are needed above 3500 feet (1067 m). There

are no "formulas" for adjusting and only repeated tests can determine the most successful adjusted recipe.

Recipe adjustments to improve baked products: This includes cake, quick breads, cookies, bar cookies, brownies, pancakes, and doughnuts. Try the lower range when a range is given.

Oven temperature—Increase 15 to 25°F (8-14°C) except for some chocolate or delicate cakes which might burn. High temperatures help to "set" the batter before the cells formed by leavening gases can expand too much. Cookies may or may not need a temperature increase.

Oven baking time—Decrease due to higher temperatures. For microwave baking, add 1-3 minutes if cake falls at end of bake time when the door is opened.

Leavening—To prevent excess rising, reduce baking powder 1/8 teaspoon at 3000 feet (0.6 ml-914 m), 1/8-1/4 teaspoon at 5000 feet (0.6–1.25 ml-1524 m), 1/4 teaspoon at 7000 feet (1.25 ml-2134 m). Reduce soda in quick breads, muffins, pancakes/waffles by 1/4 of total amount.

Sugar—For each one cup (236 ml), decrease 0-1 tablespoon at 3000 feet (0-15 ml at 914 m), 2-3 tablespoons at 5000 feet (30-45 ml at 1524 m), 4-5 tablespoons at 7000 feet (60-75 ml at 2134 m). NOTE: Fast evaporation of water leads to a higher concentration of sugar, which raises the temperature when the structure sets and may cause the cake to fall. Therefore, sugar is decreased and liquid increased. A sticky, overly browned crust and/or dip in center indicates excess sugar.

Liquid—To compensate for greater evaporation in cakes and quick breads, increase liquid. For each one cup (250 ml) add 1-2 tablespoons at 3000 feet (15-30 ml at 914 m), 2-4 tablespoons at 5000 feet (30-60 ml at 1545 m), or 3-4 tablespoons at 7000 feet (45-60 ml at 2134 m). In cookies, add 1/2-2 teaspoons (2.5-10 ml) water per recipe with the flour addition. If biscuit dough seems dry, add 1 tablespoon per one cup (15 ml/236.6 ml) flour.

Flour—Since flour strengthens the structure, for each recipe, add 1-4 tablespoons at 5000 feet (15-60 ml at 1545 m) and 2-5 tablespoons at 7000 feet (30-75 ml at 2134 m). Cookies squeezed through a press may need less flour. For self-rising flour, use only high altitude-adjusted recipes.

Eggs—Since eggs strengthen the structure and add liquid, use large eggs, although some cooks prefer extra-large eggs. For angel food, sponge or foam-type cakes or soufflés, it is important not to beat too much air (leavening) into the eggs. Use cold eggs, beat whites to soft peaks that just fold over; do not beat until stiff and dry. For foam-leavened cakes, using less sugar, more flour and a higher baking temperature also helps. For a four-egg popover recipe, add 1 more egg and decrease shortening, if listed. Above 7500 feet (2286 m) an extra egg can help the structure of rich cakes.

Fats/oils—Excess fat weakens the cell structure. For high fat-high sugar quick breads, fat may need to be reduced 2-8 tablespoons (30-120 ml) per recipe. For waffles add 1 teaspoon (5 ml) extra oil to prevent sticking.

Pans—Use larger pan size or increase number of pans. Example: For two 8-inch rounds (20 cm), use three 8-inch (20 cm) or two 9-inch (23 cm) rounds. If a large hollow air hole develops in quick breads, increase pan size or number of muffins. Generous "greasing" (1-2 tablespoons per pan—12-30 ml per pan) and flouring (if required) helps prevent sticking.

Other hints: For microwaved cakes, begin with recipes developed for microwave use. Changes may not be needed, but if the cake does not rise as high, falls, has a wet center, or is too tender, then adjustments are recommended. If center bottom is underdone, bake on inverted plate.

Sea level recipes for cookies, bar cookies and brownies may give acceptable, but different results at high altitudes. It is best to bake a test cookie to determine if more flour, liquid, or temperature changes are needed. NOTE: Cookies continue to brown a little after removal from oven, so cookies should be baked to a lighter color than desired. Since high altitude cookies are more delicate, cool 1 minute on baking sheet and carefully remove to rack. These changes may be needed for microwaved bar cookies or brownies, too.

For "mushroomed" muffins, adjust recipe for the altitude and increase number of muffins.

For pancakes, check that the insides are not doughy when the surfaces appear brown.

Commercial mixes—In cake, muffin, brownie, and other mixes, ingredients cannot be removed; therefore adjustments for conventional/microwave mixes are: adding flour, liquid, and/or an egg, plus increasing pan size or number of pans, or baking temperature/time. Follow the high altitude directions printed on the mix boxes, or contact the food manufacturer for information on adjustments.

Yeast breads—Rising time can be from 15 to 30 minutes shorter; let dough rise just until doubles. Better flavor is developed by punching down the dough twice. During rising, loosely cover dough with foil to prevent a dry crust from forming. Automatic bread machine recipes may require experimentation to prevent too dry a dough, or over-rising. Sometimes yeast is reduced and liquid added.

Pies, pastry, cheesecake and soufflés—For slightly dry pastry dough, add water by 1/2 teaspoon until the dough cleans the bowl. Thoroughly prick unfilled crusts to prevent bubbling. See: Boiled Foods: Fruits—pie fillings. Custard and pumpkin pies may need high-sided pie plates. Baked cheesecakes may have higher sides and a center area which "settles back" during cooling. Cheesecakes require much experimentation to adjust. For soufflés, see Leavened and Baked Foods: Eggs.

For questions about other commercial food products, contact the food manufacturer. They may have high altitude directions on file.

MICROWAVE COOKING BASICS

Most countertop microwave ovens have magnetron tubes that generate microwaves at a frequency of 2450 megahertz (MHz), while many combination (oven-microwave) units generate microwaves at 915 MHz. When microwaves are absorbed by food, the polar (charged) molecules such as water, fat and sugar attempt to align with the field, which changes 2450 million times per second. Heat is evolved rapidly within the food, primarily due to this molecular friction. Heating also occurs by subsequent conduction of this heat.

Cooking power for most consumer ovens is within a range of 400-720 watts; however, there are no standards. Ovens designed for both conventional and microwave cooking supply 300-400 watts cooking power. The higher the cooking power, the faster the cooking will be. The weight of food is the most accurate measure for cooking time.

Although some ovens are equipped with temperature sensing probes, most microwave heating is controlled by adjusting the time according to the cooking power, the mass and density of the food, and its starting temperature. Some ovens have variable speeds (power controls) which automatically cycle the magnetron tube off and on to vary the amount of energy used to cook foods during the specified time. The glass tray on the bottom of the countertop oven elevates food to allow microwaves to reach the bottom of the food. It should not be removed during cooking. Do not operate oven without either food or liquid in it.

Cooking power may not be evenly distributed throughout the oven; however, new features are minimizing this problem. Microwave ovens designed for cooking more than one food at a time are constructed intentionally so that different amounts of energy are directed into different areas of the oven. Manufacturers are always updating their microwave ovens and products. Refer to the manufacturer's directions for use and care of the specified oven being used.

To Estimate Cooking Power (wattage)
In 4-cup (1 l) glass measuring cup, add 2 cups (500 ml) water. Record temperature of water. Microwave on high/full power for 1 minute. Record temperature. Subtract first temperature from second temperature and multiply this by 17.5 = actual wattage. Repeat this three times and figure the average of the three. Wattage should also be listed on the manufacturer's plate on the oven.

Testing Distribution of Cooking Power
Microwaves are directed by a stirrer fan, carousel tray, or baffle, but still they may not always be directed evenly; hence, food may cook unevenly. To avoid uneven cooking, rotate dish (1/4 turn) or stir food half way through cooking time.

To test distribution of power, select test material, such as cake batter. Place an **equal amount** of test material in 9 to 12 **identical containers** such as custard cups. Space containers of test material evenly throughout the oven. Turn oven on full power for two minutes. Greater concentration of power will be indicated by visible signs of cooking in certain areas of the test product.

Water (equal amounts) may also be used for this test. The difference in water temperature from unit to unit may be determined after two minutes of heating. Higher water temperatures indicate a greater concentration of power in that area.

Mass and Shape of Food
As the mass of food increases, the length of time required to heat or cook increases. The relationship approaches linearity, but varies with each food. The major heating by microwaves penetrates only the outer 1-inch (2.5 cm) of the food. The center of large masses of food must be heated by conduction of heat from hot portions during a "standing" period following heating. "Standing time" or "carry-over cooking" allows the food to rest and continue cooking. The amount of time will vary depending on the size, volume, moisture and fat content, and density of the food. Standing time averages will usually be: 1-3 minutes for individual or small items; 5 minutes for vegetables, sauces and cakes; 10 minutes for dense main dishes; and 15 minutes for large items such as roasts and turkeys. Avoid additional cooking until standing time is completed. If necessary, add cooking time in increments of 1-2 minutes to avoid overcooking.

Techniques for Promoting Even Heating
Cooking may be assisted by subdividing food into uniform or smaller portions, rotating or stirring periodically, or periodically inverting large portions of food during cooking. Foods arranged in a circle cook more uniformly than they do in a rectangle. Avoid irregular shapes and square corners, or shield edges of foods that are thin, with a *small* piece of foil during cooking. Foil must not touch the side of the oven.

Place the thicker or more dense portions of food toward the outside of the container or arrange food in a doughnut shape. Do not overcook food. Allow for standing time after cooking. Do not remove cover during standing time.

For foods requiring long cooking times, use reduced power for part of the cooking, or manually cycle the energy on and off at intervals. The "off" interval allows time for the temperature to equalize in the food.

Initial Temperature of Food

The lower the initial temperature of the food, the longer the time required to achieve the desired end temperature by heating. Starting with frozen food increases the time even more, since additional energy is required to melt the ice (latent heat of fusion). Most foods except vegetables are thawed before cooking. Thawing may be done in the microwave oven by using reduced power prior to the actual cooking period.

Density of Food

The more dense the food, the greater the amount of microwave energy required for heating. When pieces of food differing in density or thickness are heated simultaneously, the less dense or thinner portions should be placed near the center of the container. For example, place peas in the center of a ring of mashed potatoes.

Water Content

Free water is the most polar food constituent and the most reactive to microwaves. The rapid conversion of water to steam, which occurs in microwave heating, necessitates the puncturing of food membranes or skins to prevent the food from exploding. Fats also heat very rapidly when subjected to microwaves.

When frozen foods are heated, pockets of free water can form and become superheated, resulting in an explosive release of steam. The defrost cycles on microwave ovens provide energy intermittently by pulsing on and off (cook–rest), allowing time during the off cycle for helping to equalize the temperature in the food and heating to the center of large pieces of food. Covering, during defrosting, helps to distribute the heat more evenly. Food will also need to be rotated.

Reheating Foods

Reheating foods is one of the microwave oven's greatest assets. Reheat refrigerated leftovers in serving containers, allowing 3-4 minutes/pound (0.5 kg). Frozen cooked foods will need a rest-cook cycle with rotation. To reheat foods on a dinner plate, loosely cover and allow 1-2 minutes on full power. Bread products are porous and will heat fast. A roll will heat in 10-15 seconds and may still feel cool to touch. Allow carry-over time before heating again to avoid a tough, hard roll. Slotted roasting racks are helpful to reheat bread products wrapped in a paper towel or napkin. For each cup of pasta or rice, allow 1-2 minutes to reheat.

MICROWAVES AND FOOD SYSTEMS

Meat, Poultry, Fish

Tender, boneless cuts of meat are more suitable for microwave cooking than less tender cuts. Ground beef patties, thin steaks, or chops may be cooked with high power. For roasts, reduce cooking power, particularly after the meat is hot. Reduce power during the entire cooking period to slow the process for tenderization of pot roasts. Less tender cuts can be cooked in a microwave but may require the same amount of cooking time as required for nonmicrowave cooking to tenderize collagen sufficiently. Recommendations for promoting even heating are also applicable to cooking meats by microwaves.

Roasts may need 15-30 minutes of standing time for conduction of heat to the center of the roast after microwave cooking is stopped. From 5-15 minutes of standing time is needed for fish or poultry, depending on the size. The slower the cooking and the smaller the mass, the shorter the standing time required. Since the internal temperature of the meat will rise during standing, the following internal temperatures are suggested as a guide for terminating cooking (*only* for microwave cooking):

Rare beef	125°F (52°C)
Medium beef	140°F (60°C)
Well done beef*	158°F (70°C)
Lamb	158°F (70°C)
Pork	155°F (68°C)
Reheating a fully cooked ham	131°F (55°C)
Poultry (>3.3 lb. or 1.5 kg)	170°F (77°C)

* Cook ground beef until well done.

Use only thermometers designed for microwave ovens, or use a conventional thermometer when the oven is turned off. If the meat does not achieve the desired internal temperature during the standing period, it can be sliced, covered, and cooked a few minutes longer without detrimental effects to eating quality.

An even coating of fat (about 0.5 cm thick) provides a means of basting a roast during microwave cooking. Fat that collects at the bottom of the container during roasting should be removed midway during cooking so that it will not splatter or shield the underside of the meat from microwaves.

Since less evaporation occurs than with conventional heating, only about 1/4 cup (50 ml) liquid is required for moist heat cookery of less tender cuts of meat. If a pot roast is large, cut it into chunks to lessen the time required to heat the interior to a temperature sufficient to start solubilization of collagen.

Small portions of meat, poultry or fish should be turned once during cooking. Place whole birds breast side down first and turn over midway in the cooking process. The larger the mass, the more often it should be turned. The skin of whole poultry or poultry pieces should be pierced in several places prior to cooking.

Browning aids, such as a browning utensil or coatings, may be preferred for meat patties, steaks, or chops. Coatings will be less crisp than with conventional cooking. Crisping can be achieved by placing the product under a conventional broiler after cooking. For products that tend to become dry, such as pork chops, cover with sauce.

Milk, Cheese, and Eggs

Milk tends to foam excessively when heated to the boiling point by microwaves. Therefore, a larger container is required for heating milk alone or for casseroles and other foods containing large amounts of milk. Products containing cheese should be cooked at reduced power or for a very short time to prevent a rubbery texture.

Since eggs or egg-milk combinations cook fast, the time span is reduced between optimum coagulation and over coagulation. Slow cooking and close attention are necessary. As a rule, egg foams and products composed largely of egg foam are less satisfactory when cooked by microwaves than when cooked conventionally.

Starch Pastes, Pasta, Rice

Since gelatinization of starch is enhanced by rapid heating, starch cookery is suited to microwave heating. Periodic stirring is necessary for lump free starch-thickened sauces. The sauce should be brought to a boil to achieve maximum thickening, but boiling should not be continued because the swollen starch granules may rupture.

Rice and pasta require time to absorb moisture. They cook well by microwaves, but little time is saved by this method of cookery. Additional time, at reduced power, will be necessary after the water has been brought to a boil. Adding a small amount of oil reduces the tendency to boil over. Pasta, rice or cereal may be cooked for individual servings or in quantity. Cook in a dish large enough to boil water without boiling over.

Sugar Solutions

Microwave cooking of sugar solutions minimizes the problems of scorching and of crystal build-up on the sides of the cooking container. With candy, the solution can be brought to a rapid boil, and boiling can be continued at a reduced level of microwave power. A large container is essential to prevent boiling over.

Jams or preserves can be made in small batches by microwave heating. The power should be reduced after the mixture has reached the boiling point to minimize the possibility of overcooking.

Vegetables and Fruits

Vegetables and fruits are easily cooked by microwaves. The speed of cooking favors the retention of color and pleasing flavor. Overcooking may result in a rubbery texture. For vegetables which require short cooking by conventional methods, little time is saved by microwave heating. However, for long-cooking vegetables, such as whole potatoes, time may be reduced as much as 85 or 90 percent. Many vegetables can be cooked without added water, but for those with a high cellulose content, texture is improved by cooking with a small amount of water. Most vegetables and fruits should be covered. Skins on whole vegetables serve as a cover and should be pierced to permit steam to escape and to minimize the possibility of explosions.

Baked Products

Cakes, muffins and other quick breads cook well by microwaves, but do not form brown crusts. Leavening should be reduced by one-third for microwave cooking. For products prepared from mixes, the texture may be improved by allowing the batter to stand for 15-20 minutes before cooking. The surface of the products will appear slightly moist when cooking is complete. Allow 10-12 minutes standing time for layer cakes so that some cooking will continue for a short time. Lining the container with waxed paper will reduce the tendency for sogginess on the bottom of the cake. Biscuits, yeast bread, cookies (other than bar cookies), and cakes without shortening are less satisfactory when cooked by microwaves than when baked conventionally.

Fats and Oils

Fats and oils heat rapidly when exposed to microwaves. However, they do not heat uniformly, which can result in very dangerous spattering of hot fat and uneven cooking. Therefore, deep fat frying in a microwave is not recommended.

Baby Food

A 4 1/2 ounce (128 g) jar of baby food may be heated with the lid off, for 45-90 seconds. To heat a baby bottle, place the bottle and 1/2 cup (125 ml) of room temperature water in oven. Heat on high for 15 seconds. Refrigerated milk or formula requires 30-40 seconds on high. Allow at least one-minute standing time before stirring or shaking to help avoid having some of the formula dangerously hot. Check temperature before serving to prevent possible burning of baby.

PREPARATION OF MEALS BY MICROWAVES

Heat long-cooking foods and those which hold heat well first. Allow for continued cooking (particularly in large masses of food) after the microwave heating has stopped. If more than one food is cooked at one time, extend the length of time. Cooking time must be adjusted according to the amount of food cooked at one time. Use a microwave oven for partial cooking of some foods. Take advantage of this appliance for fast defrosting and for fast reheating of foods.

Plan meals for the characteristics of microwave heating, such as less evaporation of liquid, greater expansion of leavened products, rapid evolution of steam, and continued heating after oven is turned off, particularly in dense foods.

Remember that casseroles and entrees will require only about one-fourth to one-third as long for microwave as for conventional cooking. An interruption for testing doneness will not be detrimental to the finished quality of any product.

Adapting Conventional Recipes

Use successful microwave recipes as a guide to modify conventional recipes. Consider the major ingredients and their suitability for microwave cooking. Use small recipes first, as they are easier to convert than large recipes.

Recipes with a large amount of moisture may need adjusting, as the microwave does not evaporate as much liquid. A general rule is to reduce the liquid in the conventional recipe by one-fourth. To determine cooking time, consider that 1 cup (250 ml) of tap water boils and 1 cup of food heats in 1 1/2 to 3 minutes on high. If your oven takes longer, add additional time to recipes. To be more accurate, weigh food and estimate 6-7 minutes/pound (0.5 kg) on high, depending on oven wattage. Increase cooking time as the amount of food increases. Other considerations are shape, volume, weight, density, initial temperature, oven wattage, peak electric use hours and the cooking container. Undercook when experimenting with new foods and recipes and then continue to cook if more time is judged to be needed.

High Altitude

Adjust conventional recipes by increasing flour 1/4 cup (60 ml) and liquid by 1/3 cup (80 ml). Since boiling occurs at a lower temperature, it may take longer (1-2 minutes) to cook dense foods.

Choosing Utensils

Choose utensils that are transparent to microwaves, such as glass, paper, dishwasher-safe plastics, china (without metal rim), earthenware, or straw. Do not use metal, aluminum foil, paper goods made from recycled paper, or

Melamine ware. Plastic containers may deform from heat if used for syrups or high fat foods. Plastic film should be vented when used as a cover. Slit boil-in-bag containers to permit steam to escape.

Test dishes for microwave use by placing one cup (250 ml) of tap water and an empty dish in oven. Heat on high for one minute. If the water gets hot and the dish does not, it can be used. Use only glass-ceramic dishes if oven has infrared browner. Metal reflects microwaves; if recommended by manufacturer, foil may be used for shielding.

Use only thermometers designed for microwave ovens. However, conventional thermometers can be used to test the temperature of food when the oven is turned off.

As covers, use casserole lids, plastic film, paper plates, waxed paper, paper towels, or napkins. Covering food traps heat and speeds up the cooking time. Choose porous materials, such as paper towels, if the surface of the food is dry. If possible, use round containers for cooking instead of those with square corners. A heatproof glass may be used in the center of casseroles to achieve the doughnut shape. Special microwave browning dishes are heated empty in the microwave to absorb energy. The bottom of the dish absorbs energy instead of transmitting it, thus causing the food to brown when placed on the very hot surface.

Reducing the Speed of Cooking
To retard the rate of cooking, utilize oven controls for reducing power, cycle energy on and off manually, or introduce an additional load, such as a container of water to absorb part of the energy.

Safety
Do not melt paraffin; it does not have any moisture to absorb energy. Home canning is not recommended because there is no way of knowing how much pressure is building inside the sealed jar. Home canned vegetables and meats should be processed conventionally to ensure destruction of all *Clostridium botulism* spores.

Cooking sponge cakes, deep-fried foods, popovers, hard- and soft-cooked eggs in the shell are not recommended. Follow manufacturer's recommendations for cooking popcorn.

Do not operate a microwave oven empty. If the cavity does not have enough food mass, the microwaves may bounce back to the magnetron tube and shorten the life of the oven. Cooking containers will become hot to the level of food in the container as heat in the food transfers to the container. Use hot pads as a precaution. Transfer frozen, prepackaged convenience and frozen prepared food to nonmetal containers for cooking.

To clean a microwave oven, heat 1 cup (250 ml) water until it steams. Wipe inside of cavity with soft cloth after steam softens food. Microwave ovens are always being updated, hence the need to follow manufacturer's direction for use and care of oven. To remove food odors, add lemon juice or baking soda to warm water, close oven door and let mixture sit for a short time.

Enhancing Appearance of Food
The speed of microwave cooking and the tendency for food to give up heat to the surrounding atmosphere are not conducive to browning and crusting of the surfaces of food. Browning occurs if the food requires cooking for 20 minutes or more. Many foods are acceptable without a brown crust. Browning ingredients and sauces may be used to advantage on other foods.

To Enhance Food Appearance

Biscuits	Brown in browning utensil. Top with toasted sesame seeds, poppy seeds, cinnamon and sugar, or dry onion mix.
Cakes	Sprinkle brown sugar and cinnamon in container before pouring batter, or sprinkle on top after cooking. Use topping for upside-down cakes. Cool baked cake and then ice it.
Casseroles	Top with toasted bread crumbs, crushed chips, cereals, toasted nuts, grated cheese, chopped bacon. Brown under broiler before or after cooking by microwave.
Meat, Poultry	Brush with brown sauce or gravy mix. Sprinkle with paprika or dry onion soup.

Fish   Top, or coat, with toasted bread crumbs, nuts or grated cheese. Cover with sauce. Glaze with brown sugar or other ingredients. Brown on grill or broil and complete cooking by microwave. Use browning utensil.

Muffins   Include bran as an ingredient.

Pastry   Prepare with milk instead of water. Add a few drops of food coloring.

SLOW-COOKER COOKING

The slow cooker is an electrical appliance, specifically designed to cook foods slowly, generally taking 4 to 12 hours. There are at least two general types of slow cookers. One has a stoneware or crockery container in an electrical heating base. Heating elements often wrap around the sides of the stoneware and control settings may include low and high. Another type of slow cooker consists of a metal container that is heated from the bottom. The temperature control may include a variety of temperature settings, but a range for slow-cooking will be specifically noted. Variations of these two types, such as a crockery bowl in a metal cooking pot with an adjustable thermostat, are also available.

Check the manufacturer's instructions for information on using a specific slow cooker, and note the recommended cooking times and settings. Always cook for the suggested time on the proper setting.

The slow cooker is helpful in cooking less tender cuts of meat, soups, and stews. A variety of other dishes, including casseroles, vegetables, beans, hot drinks and cereals can also be prepared in a slow cooker. For cooks who wish to adapt favorite recipes to the slow cooker, this information provides a guide.

Adapting Recipes
- Look for steps that may be omitted (for example, browning aspects). Most ingredients can be added at the beginning of the cooking period.
- For most recipes, one cup of liquid is sufficient, but a general rule of thumb is to use only one-half as much as in a standard recipe. Additional amounts may be

needed if cooking a product such as pasta or rice, which absorbs liquid.
- Crushed or ground seasonings should be added near the end of cooking because they become too mild during the long cooking period. Whole seasonings may withstand the long cooking, and their flavor may intensify, so use about one-half as much as usual.
- Rice and pasta become gummy with long cooking; add them during the last hour of cooking. If pasta is to be cooked before adding to the slow cooker, cook the pasta just until al dente.
- Sour cream and milk curdle during long cooking and should be added during the last 30 minutes. Condensed soups are satisfactory for the longer cooking times and may be substituted for milk or cream.

Food Placement
Placement of foods in the cooker may determine the success or failure of the dish. Vegetables such as potatoes, carrots, and onions cook very slowly and should be kept small to medium in size and placed in the bottom of the pot. This also allows the meat juices to drip over the vegetables, enhancing their flavor. Unless the meat is very fatty, a rack usually is not necessary. If a sauce is used over meat or vegetables, it should be poured evenly over the food.

Timing Factors
- Allow plenty of cooking time, especially on low. Remember that timing is generally not critical and an extra hour will not greatly affect the food. Always cook the food until it is completely done and tender.
- The temperature of the food when placed in the cooker will affect the cooking time. Most cooking times are based on using refrigerated foods. Frozen foods require a longer cooking time than refrigerated foods. In some slow cookers, frozen foods are not recommended or specific instructions must be followed; check the use and care manual for information on frozen foods in a specific slow cooker.
- The volume of food will affect the cooking time. Refer to the use and care manual as a guide for the amount of food suited to the size of cooker you are using. Generally the cooking times are based on units one-half

to three-fourths full. Do not overfill slow cookers. If cooking a very small amount of food in a cooker with a wrap-around heating element, the cooking time may be shortened, and the cooking progress should be checked.

- If there is much fat present, meat should first be trimmed, browned, or broiled to eliminate as much fat as possible.

- Minerals (as in some hard water areas) and acids increase the cooking time in dried legumes and may affect other foods in hard-water areas. Sugars and acid foods such as tomatoes and molasses should be added only after the legumes have been softened completely.

- Some foods that should be added during the last 1 to 2 hours of cooking including seafood, canned or frozen vegetables, and fresh mushrooms. Frozen vegetables and seafood should be thawed before being added to the cooker.

The following chart provides guidelines for cooking adapted recipes in a slow cooker.

Time Guide for Slower Cooker[a]

	Time in Slow Cooker	
If Recipe Says:	High Setting	Low Setting
15 to 30 min.	1.5 to 2.5 hrs.	4 to 8 hrs.
35 to 45 min.	3 to 4 hrs.	6 to 10 hrs
50 min. to 3 hrs.	4 to 6 hrs.	8 to 18 hrs.

[a] Courtesy of Rival Manufacturing Company

CONVECTION OVEN COOKING

A convection oven uses forced hot air to speed the baking/roasting process, thus saving time in the oven.

Types of Ovens
Convection ovens are available as countertop, wall, or range ovens. The three different convection modes are: convection with heating element located around the fan and outside the oven cavity; convection bake with heating elements on inside oven cavity and a fan to circulate the hot air; and microwave/convection—a combination

of microwave power and heating element and fan located outside the oven cavity.

Factors to consider when adapting conventional recipes to the convection oven include time, temperature settings, and type of bakeware. Check the appliance manual for specific suggestions.

Baking
Temperature—Oven Temperature is decreased by 25-50°F (15-30°C). Bakes items with sufficiently brown surface but undercooked centers may need the temperature lowered an additional 25°F (15°C) and increased baking time.

Time—Baking time varies with moisture content, richness, and depth of product. Check times in appliance manual. Generally for convection or convection bake modes, times can be reduced as follows: thin items—about 1-2 minutes; cakes—5 minutes; and large turkeys—up to 30 percent. Preheating is necessary only for food baked less than 20 minutes and products such as popovers where steam rising is important.

Size—Large items may not cook fully in the center. Divide these into small units such as two or three loaves instead of one.

Roasting:
Temperature—Use the same temperature as conventional recipes. Preheating is not necessary. Roast meats uncovered except turkey, which should be covered once it is golden brown.

Time—Reduce the time by one-third depending on the size of product. If meat is too rare in the center, lower the temperature and increase the time.

Controlling Moisture—Baste roasts with glaze or barbecue sauce only during the last 10 to 20 minutes; use a drip pan with 1/4 inch (0.8 cm) water. Frozen meat can be roasted in the same time as thawed meat in a conventional oven.

Quantity Cooking
A major benefit of convection ovens is quantity cooking. It is possible to bake three racks of cookies, biscuits, pizzas or similar items at the same time and have even browning of the product.

Bakeware

Metal bakeware such as aluminum and heavy-duty aluminum foil gives the greatest time savings, followed by cast iron and black steel. Ceramic and glass dishes require conventional baking time. Baking sheets should have only one edge. Shallow pans that allow at least 1 to 1 1/2 inches (2.5 to 3.7 cm) of space between the bakeware and the oven wall are recommended for roasting. Covered casseroles can be used in the convection oven, but the convection benefit of reduced time is minimized. Cookware should be centered in front of the fan with 1 inch (2.5 cm) air space on all sides.

FOOD LABELING

Federal laws governing food labeling are the concern of the following agencies:

FDA—Food and Drug Administration. A federal agency empowered to enforce five specific laws relating to checking all foods and food additives for safety, purity, and wholesomeness (Federal Food, Drug and Cosmetic act; Fair Packaging and Labeling Act; Radiation Control and Health Safety Act; Public Health Service Act; and Nutrition Labeling and Education Act of 1990). FDA officers inspect plants where food is processed. The agency is also responsible for the accuracy and correctness of information of packages and labels on the majority of foods.

USDA—U.S. Department of Agriculture. This department is responsible for meat and poultry safety, quality, and labeling (see FSIS). The Research Service of this agency conducts food and nutrition research. Areas investigated include effects of processing and cooking on nutrient content of food and food preferences and habits of individuals and groups. Officers inspect plants for the cleanliness of the plant and workers and health of the animals slaughtered.

FSIS—Food Safety and Inspection Service. This agency of USDA inspects all red meats and poultry for purity, wholesomeness, and truthful labeling. This agency monitors meat and poultry for hazardous residues above Environmental Protection Agency (EPA) levels.

AMS—Agricultural Marketing Service. It is an agency within USDA which grades fresh, frozen, canned and dehydrated vegetables into levels of quality. It also establishes grade standards for most fresh and processed fruits and for fresh meat (p75), poultry (p92), and eggs (p96).

USDC NMFS—U.S. Department of Commerce–National Marine Fisheries Service. This agency provides voluntary inspection of fish and fish products. The service develops standards and specifications for quality, conditions, quantity, grade and packaging of fish, processing plants and fish products.

USPHS—U.S. Public Health Service. This agency safeguards the purity of dairy products. Evidence of contamination is monitored according to standards stated in the Grade A Pasteurized Milk Ordinance. This agency assists with enforcement of standards relating to sanitation, taste, odor, and milkfat content. The federal code defines milk fat standards for various milk products. This agency works with USDA in monitoring processing at milk plants.

Ingredient Labeling

For nonmeat and nondairy products, the ingredient label is regulated by FDA. Meat and dairy product labels are regulated by USDA. The ingredient label must tell the name of a product, the form it is in, and identify anything which has been added to it. Ingredients must be listed by weight, from highest weight to lowest content.

The Nutrition Labeling and Education Act of 1990 established requirements for the declaration of all ingredients in standardized foods, declaration of certified color additives in foods, declaration of sources of protein hydrolysates, and declaration on the nutrition label of percentage of a fruit or vegetable juice in a beverage claiming to contain such a juice. The effective date for total percentage juice labeling, and ingredient labeling of standardized foods and certified colors was May 8, 1993.

Nutrition Labeling

The Nutrition Labeling and Education Act (NLEA) of 1990 brought about a great many changes to the labeling practices monitored by the Food and Drug Administration. Proposed rules for implementing the act were reviewed throughout 1992, with final rules published in the Federal Register on January 6, 1993. Effective date for mandatory nutrition labeling using the specified format was May 8, 1994.

The purpose of the labels is to allow comparison of foods for nutritional content. The 1990 law requires mandatory nutritional labeling of most foods offered for sale and regulated by FDA. The nutrition label is required to include information on total calories and calories from fat and on amounts of total fat, saturated fat, cholesterol, sodium, total carbohydrates, dietary fiber, sugars, protein, vitamin A, vitamin C, calcium and iron—in that order. Manufacturers also may voluntarily declare information on calories from saturated fat and on amounts of polyunsaturated and monounsaturated fat, soluble and insoluble fiber, sugar alcohol, other carbohydrate, potassium, additional vitamins and minerals for which Reference Daily Intakes (RDIs) have been established, and the percent of vitamin A present as beta-carotene. The information presented on the label is to represent the packaged product prior to consumer preparation.

The standard format for nutrition information on food labels consists of:

1. the quantitative amount per serving of each nutrient except vitamins and minerals;
2. the amount of each listed nutrient as a percent of the Daily Value for a 2,000 kilocalorie diet;
3. a footnote with reference values for selected nutrients based on 2,000 kilocalorie and 2,500 kilocalorie diets; and
4. kilocaloric conversion information.

Voluntary labeling applies to fresh produce and seafood unpackaged or packaged at retail. Nutrition information at point of purchase may be presented via large placards, brochures or videos. Unless claims are made about the following nutrients, labeling is voluntary: kilocalories from saturated and unsaturated fat, total carbohydrates, protein; unsaturated fat or amounts of polyunsaturated and monounsaturated fats (g); insoluble and soluble fiber; protein as percentage of RDI for foods other than infant foods; potassium (mg), thiamin, riboflavin, niacin and other vitamins and minerals as percentage of RDI unless added as a supplement. Nutrition information is given "as con-sumed" for raw fish and "as packaged" for raw fruits and vegetables. Sixty percent of retailers surveyed must be in compliance for at least 90 percent of the 20 most frequently consumed raw fruits, raw vegetables, and raw fish.

The USDA-Food Safety and Inspection Service (FSIS) also published in the January 6, 1993, issue of the *Federal Register* final rules amending the Federal meat and poultry products inspection regulations permitting voluntary nutrition labeling on single-ingredient, raw meat and poultry products, and mandatory nutrition labeling for all other meat and poultry products. The FSIS rules were an effort to parallel as much as possible the nutrition labeling regulations of the NLEA. The effective date for USDA nutrition labeling of meat and poultry products was July 6, 1994.

Nutrition Profile

The Food and Drug Administration (FDA) regulations established Reference Daily Intakes (RDIs) and Daily Reference Values (DRVs). The final rules published in 1993 retained the values established by FDA in 1973 for vitamins and minerals but changed the terms for those values from U.S. Recommended Daily Allowances (U.S.RDA) to Reference Daily Intake (RDI). The regulation also established label reference values for eight other nutrients, including fat, cholesterol, and fiber. The values were established as Daily Reference Values (DRVs). To avoid confusion, all reference values on food labels are referred to as Daily Values or DVs.

Reference Serving Sizes

Serving sizes are defined as the amount of food customarily consumed per eating occasion. Nutrient content values listed on product labels are then based on the standardized serving sizes. A total of 184 food product categories have been defined—139 by FDA, 23 for meats by FSIS, and 22 for poultry by FSIS. Additionally, there are specific rules for single-serving containers and items that come in discrete units such as slices of bread or individual muffins or cookies.

Descriptive Terms

FDA has defined the terms "free," "low," "light" or "lite," "reduced," "less," and "high," as well as "good source," "very low" (for sodium only), "lean," "extra lean," "fewer," "more," and "added" (for "fortified" or "enriched").

Some of the specific definitions are as follows: Criteria for the term "**low**" are based on reference serving sizes or for products having reference amounts of less than 30 g or 2 tablespoons, per 50 g, and thus **low fat** is a food with not more than 3 g of fat, **low calorie** is a food with not more than 40 calories, and **low sodium** is a food with not more than 140 mg sodium. **Light (lite)** can be used if the food has one-third fewer kilocalories than a comparable product, if half of the kilocalories are from fat and the fat has been reduced by 50 percent; and **cholesterol-free** claims may be made only on foods that contain 2 g or less of saturated fat per reference serving.

The terms "**high**" and "**good source**" are based on a percentage of the Daily Value (DV) of the specific nutrients in a reference serving. "**High**" is defined as 20 percent or more of the DV, and "good source" as between 10 and 19 percent of the DV. See Table 1.2 for a more complete list of nutrient content labeling claims definitions.

Health Claims

FDA allows food manufacturers to make unqualified health claims on food labels based on "significant scientific agreement" or an "authoritative statement of a scientific body of the U.S. government or the National Academy of Sciences." (See www.cfsan.fda.gov for more information.) FDA also permits other qualified health claims when there is emerging evidence of a relationship between a food or dietary supplement and reduced risk of a disease or health-related condition, structure and function claims, and statements of dietary guidance that are truthful and non-misleading. Health claims link various nutrients (i.e., dietary fat, sodium, and calcium) and food groups (i.e., fruits and vegetables and fiber-containing grain products) to reduction in risk of diseases, such as coronary heart disease, cancer, hypertension, osteoporosis, and other diseases. Structure/function claims that describe the role of a nutrient or dietary ingredient in affecting structure or function in the human body are also allowed by FDA. Some examples of various types of claims are as follows:

Table 1.2

Examples of Health Claims for Nutrition Labels

Type of Claim	Example Food	Food	Model Claim
Unqualified	Sodium and hypertension	Low sodium	Diets low in sodium may reduce risk of high blood pressure, a disease associated with many factors.
Unqualified	Dietary fat and cancer	Low fat	Development of cancer depends on many factors. A diet low in total fat may reduce the risk of some cancers.
Qualified	Walnuts and heart disease	Whole or chopped walnuts	Supportive but not conclusive research shows that eating 1.5 ounces per day of walnuts, as part of a low saturated fat and low cholesterol diet and not resulting in increased caloric intake, may reduce the risk of coronary heart disease.
Structure/Function	Calcium	Dairy foods	Calcium builds strong bones.

Trans Fat Content

Beginning January 2006, food manufacturers were required by FDA to list trans fat content of foods in grams separately on the Nutrition Facts label in two circumstances:

1. the food contains over 0.5 g trans fat per serving.

2. a health claim is made about fat, fatty acids, or cholesterol.

The requirement to include trans fat on the label is based on emerging scientific evidence linking consumption of trans fats to increased risk of coronary heart disease. FDA has not yet established a Daily Value for trans fats. Consumers are advised to use the food label information as a guide for choosing foods low in saturated fat, trans fat, and cholesterol.

Food Allergens

The Allergen Labeling and Consumer Protection Act, which became effective in January 2006, is designed to provide specific information to consumers regarding major food allergens contained in foods. Foods with any ingredient containing egg, wheat, peanuts, milk, tree nuts, soy, fish, or crustaceans (shellfish) are required to state the common name of the food in parentheses or an additional statement in the ingredient list on the product label.

Nutrient Content Claims and Standardized Foods

The large number of substitute products now in the marketplace has created a need for standardization of the substitutes. The substitute product cannot be nutritionally inferior to the original product. The substitute product must also have similar performance characteristics to the original product, and if it does not, this must be clearly printed on the label. The product must contain the ingredients used in the original product, but may also contain certain safe and suitable ingredients, provided the product is not nutritionally and functionally inferior to the original food in its performance characteristics.

Exemptions from Mandatory Programs

Foods exempt from mandatory nutrition labeling requirements include:

1. foods offered for sale by small businesses;
2. food sold in restaurants or other establishments in which food is served for immediate human consumption;
3. foods similar to restaurant foods that are ready to eat but are not for immediate consumption, are primarily prepared on site, and are not offered for sale outside of that location;
4. foods that contain insignificant amounts of all nutrients subject to the law such as coffee and tea;
5. dietary supplements, except those in conventional food form;
6. infant formula;
7. medical foods;
8. custom-processed fish or game meats;
9. foods shipped in bulk form; and
10. donated foods.

Otherwise exempted foods that make a nutrient content claim or health claim forfeit the exemption.

FOOD ADDITIVES

By the broadest definition, a food additive is any substance that becomes a part of the food product when added directly or indirectly.

The 1906 Food and Drug Act and the more comprehensive Food, Drug and Cosmetic Act of 1938 gave the government authority to remove adulterated and obviously poisonous foods from the market. The 1958 Food Additives Amendment, The Delaney Clause, and the 1960 Color Additives Amendment are laws specifically set up to regulate additives in foods. In these laws, legislation shifted the burden from the government to the manufacturer to prove that a food additive is safe. FDA was authorized to regulate additives only on the basis of safety. The agency has no power to limit the number of additives approved, or to judge whether an additive is really needed.

Under the Food Additives Amendment, however, two major categories of additives were exempted initially from the testing and approval process. The first group of some 700 substances belongs to the "generally recognized as safe" (GRAS) list, which includes salt and other common ingredients such as sugar and many herbs and spices, considered safe based on either a history of safe use before the Food Additives Amendment was passed in 1958 or the results of scientific studies. The GRAS list was designed to allow continued use of substances considered harmless because of past extensive use without known harmful effects. However, GRAS additives have been undergoing systematic testing for safety since 1971. Also exempted from testing were the "prior sanctioned substances," those that had been approved before 1958 for use in food by either FDA or USDA.

TABLE 1.3

Nutrient Content Labeling Claims Definitions

Descriptive Term	Definition by Food Energy/Nutrient Category						Comments
	Calories	**Fat**	**Saturated Fat**	**Cholesterol**	**Sodium**	**Sugars**	
Free	Contains no amount of, or only trivial amounts of, any one or more of the categories: calories, fat, saturated fat, cholesterol, sodium, and sugars.						**Synonyms: "no," "without," "zero"**
	<5 kcal/ serving	<0.5 g/ serving	<0.5 g/ serving	< 2 mg/ serving	< 5 mg/ serving	< 0.5 g/ serving	
Low	40 kcal or less/serving	3 g or less/ serving	1 g or less/ serving	20 mg or less/serving	140 mg or less/serving	Not defined	**Synonyms: "little" ("few" for calories). Amount applies/50 g food if references amount 30 g or less (small serving)**
Very low					35 mg or less/serving		
Lean		<10 g/ serving	< 4 g/ serving	<95 mg/ serving			**Amounts apply per 100 g also**
Extra lean		< 5 g/ serving	<2 g/ serving	< 95 mg/ serving			**Amounts apply per 100 g also**
Light or Lite	Contains 1/3 fewer calories than ref. Food	Contains 1/2 fat of ref. Food			Content reduced by 50% in low-calorie/ low-fat food		**If food gets 50% or more of calories from fat, must have 50% reduction in fat**
Reduced	Nutritionally altered product has at least 25 percent less of nutrient or calories than reference product. Claim can't be made if reference food meets requirement for "low" claim.						**Synonyms: "lower" ("fewer" for calories)**
Less	Food, altered or not, has 25 percent less of a nutrient or calories per reference serving.						**Synonym: "fewer"**
High	Contains 20 percent or more of DV for nutrient per reference serving.						
More	Food, altered or not, has 10 percent or more of DV of nutrient preference serving.						
Good Source	10-19 percent of DV of a particular nutrient preference serving.						
Fresh	May only be used on a food that is raw, never frozen or heated and contains no preservatives.						

An additive is intentionally incorporated into foods for one or more of these four purposes:

- to maintain or improve nutritional value
- to maintain freshness
- to help in processing or preparation
- to make food more appealing.

Some people have adverse reactions to certain foods and some food additives. When that happens, reactions can range from headaches or hives to seizures or death. If you suspect an allergic reaction to an additive, contact your physician for diagnosis and treatment.

The following Additives Index is a guide to some substances commonly added to foods:

ADDITIVES INDEX

Key Definitions

Nutrients:	enrich (replace vitamins and minerals lost in processing) or fortify (add nutrients that may be lacking in the diet).
Preservatives (Antimicrobials):	prevent food spoilage from bacteria, molds, fungi and yeast; extend shelflife; or protect natural color or flavor.
Antioxidants:	delay/prevent rancidity or enzymatic browning.
Emulsifiers:	help to distribute evenly tiny particles of one liquid into another; improve homogeneity, consistency, stability, or texture.
Stabilizers, Thickeners, Texturizers:	impart body; improve consistency or texture; stabilize emulsions; affect mouth feel.
Leavening Agents:	affect cooking results—texture and volume.
pH Control Agents:	change/maintain acidity or alkalinity.
Humectants:	cause moisture retention.
Maturing and Bleaching Agents, Dough Conditioners:	accelerate the flour aging process; improve baking qualities of flour.
Anticaking Agents:	prevent caking, lumping or clustering of a finely powdered or crystalline substance.
Flavor Enhancers:	supplement, magnify or modify the original taste and/or aroma of food without imparting a characteristic flavor of their own.
Flavors:	heighten natural flavor; restore flavors lost in processing.
Colors:	give desired, appetizing or characteristic color of food.
Sweeteners:	make the taste of food more agreeable or pleasurable.
Alternative Sweeteners:	make the taste of food sweeter with no or few added calories.
Fat Substitutes:	affect the mouth feel of foods by replacing fat with fat substitutes.

Additives

Key to Abbreviations used in Following List:

Stabil-thick-tex	=	stabilizers-thickeners-texturizers
Leavening	=	leavening agents
pH control	=	pH control agents
Mat-bleach-condit	=	maturing & bleaching agents, dough conditioners
Anticaking	=	anticaking agents

Acetic acid (vinegar) . . . . pH control
Acetone Peroxide. . . . . mat-bleach-condit
Acesulfame-K . . . . sweetener
Agar . . . . stabil-thick-tex
Alginate . . . . stabil-thick-tex
Aluminum, sodium sulfate . . . . pH control
                                 firming agent
Ammonium alginate. . . . . pH control
Annatto extract . . . . color
Arabingalactan (larch gum) . . . . . . stabil-thick-tex
Ascorbic acid (Vitamin C) . . . . nutrient, antioxidant
Aspartame . . . . sweetener
Beet powder . . . . color
Benzoyl peroxide . . . . mat-bleach-condit
Benzoic acid . . . . preservative
   (sodium benzoate)
Beta-carotene. . . . . nutrient, color
BHA (butylated hydroxyanisole). . . antioxidant
BHT (butylated hydroxytoluene). . . antioxidant
Calcium bromate. . . . . mat-bleach-condit
Calcium carbonate . . . . anticaking
Calcium caseinate . . . . stabil-thick-tex
                                 nutrient (fortify)
Calcium phosphate . . . . leavening
   (monocalcium phosphate)   pH control
                                 nutrient
Calcium propionate . . . . preservative
Carob bean gum . . . . stabil-thick-tex
Caramel. . . . . color
Carrageenan . . . . stabil-thick-tex
Carrot oil. . . . . color
Cellulose . . . . stabil-thick-tex
Chlorine & chlorine dioxide . . . . . . mat-bleach-condit

Citric acid . . . . pH control,
                                 antioxidant
Cyclamate . . . . sweetener
                                 (legal in Canada,
                                 illegal in USA)
Diglycerides . . . . emulsifier
DSG (disodium 5'-guanylate) . . . . . flavor enhancer
EDTA . . . . antioxidant
   (ethylene-diaminetetra-acetate)
Ethyl vanillin . . . . flavor
Ferrous gluconate. . . . . nutrient, color
Ferrous sulfate (iron) . . . . nutrient
FD & C Colors . . . . color
   Blue 1
   Blue 2
   Green 3
   Orange B
   Citrus Red
   Citrus Red 2
   Red 3 (provisional)
   Red 40
   Yellow 5
   Yellow 6
Gelatin. . . . . stabil-thick-tex
Glutamic acid. . . . . flavor
Glycerine. . . . . humectant
Grape skin extract . . . . color
Guar gum . . . . stabil-thick-tex
Gum arabic (acacia) . . . . stabil-thick-tex
Heptyl paraben. . . . . preservative
Hydrogen peroxide . . . . mat-bleach-condit
HVP . . . . flavor enhancer
   (hydrolyzed vegetable protein)
Iron-ammonium citrate. . . . . anticaking
Iron, reduced . . . . nutrient
Isopropyl citrate. . . . . antioxidant
Lactic acid. . . . . pH control
                                 preservative
Lecithin. . . . . emulsifer
Limonene . . . . antioxidant, flavor
Locust bean gum. . . . . stabil-thick-tex
   (carob bean)
Magnesium silicate . . . . anticaking
Mannitol . . . . sweetner

Methyl cellulose................ anticaking
                              stabil-thick-tex
                              bulking
Methyl paraben .... preservative
Modified food starch............ stabil-thick-tex
Monoammonium glutamate...... flavor enhancer
Monoglyceride .... emulsifier
Monosodium Glutamate (MSG)... flavor enhancer
Niacinamide .... nutrient
Nitrates .... preservative
Oleoresins..................... flavor
Parabens .... preservative
Pectin........................ stabil-thick-tex
Phosphoric acid................ pH control
Polysorbates (60, 65, 80) .... emulsifiers
Potassium bromate............. mat-bleach-condit
Potassium iodide (iodine) .... nutrient
Proprionic acid & salts........... preservative
Propylene glycol .... stabil-thick-tex
Propyl gallate.................. antioxidant
Riboflavin .... nutrient, color
Saccharin .... sweetner
Saffron....................... color
Silicon dioxide................. anticaking
Simplesse® .... fat substitute
Sodium ascorbate (Vitamin C) .... nutrient
Sodium citrate................. pH control
Sodium erythorbate .... antioxidant

Sodium proprionate .... preservative
Sodium stearyl fumarate......... mat-bleach-condit
Sorbic acid .... preservative
Sorbitan monostearate.......... emulsifier
Sorbitol .... flavoring, humectant
Stearyl lactylate .... emulsifier
                              conditioner
                              whipping agent
Sugar (fructose, invert sugar,
    glucose, sucrose, dextrose)..... sweeteners
Sulfite salts................... preservative
Sulfur dioxide.................. preservative
Tartaric acid (cream of tartar) .... pH control
TBHQ .... antioxidant
    (tertiary butyl hydro-quinone)
Thiamin....................... nutrient
Titanium dioxide .... color
Tocopherols (Vitamin E) .... nutrient, antioxidant
Tragacanth gum................. stabil-thick-tex
Trisodium phosphate .... stabil-thick-tex
Turmeric .... color
Vanilla .... flavor
Vanillin .... flavor
Vitamin A .... nutrient

SUPPLEMENTARY AIDS TO FOOD PREPARATION

Baking Temperatures and Times

Type of Product	Oven Temperatures °F	(°C)	Baking Time (minutes)
Breads			
Biscuits	425 - 450	(220 - 230)	10 - 15
Corn bread	400 - 425	(200 - 220)	30 - 40
Cream puffs	425 - 450	(220 - 230)	45 - 60
Muffins	400 - 425	(200 - 220)	20 - 25
Popovers	350 - 375	(180 - 190)	
Quick loaf breads	350 - 375	(180 - 190)	60 - 75
Yeast bread	400	(200)	30 - 40
Yeast rolls, plain	400 - 425	(200 - 220)	15 - 25
sweet	375	(190)	20 - 30
Cakes with fat			
Bundt	350	(180)	40 - 45
Cup, 12 cups	350 - 375	(180 - 190)	15 - 25
Layer, 9 inch	350 - 375	(180 - 190)	20 - 25
Loaf	350	(180)	45 - 60
Cakes without fat			
Angle food and sponge	350 - 375	(180 - 190)	30 - 45
Jelly roll	325	(160)	22 - 28
Cookies			
Drop	350 - 400	(180 - 190)	8 - 15
Rolled	375	(190)	8 - 15
Meringues	200	(100)	60 + 120 after oven off
Egg, meat, milk, and cheese dishes			
Cheese soufflé, custards			
(baked in a pan of hot water)	350	(180)	45 - 60
Macaroni and cheese	350	(180)	25 - 30
Meat loaf	350	(150)	60 - 90
Rice pudding (raw rice)	300	(150)	120 - 180
Scalloped potatoes	350	(180)	60 - 75
Pastry			
One-crust pie (custard type),			
unbaked shell	400 - 425	(200 - 220)	30 - 40
Meringue on warm cooked filling			
in prebaked shell	350	(180)	12 - 15
Shell only	450	(230)	10 - 12
Two crust pies			
with uncooked filling	400 - 425	(200 - 220)	45 - 55
with cooked filling	425 - 450	(220 - 230)	30 - 45

* When baking in ovenproof glassware, reduce oven temperature by 25°F or by 10°C.

NOTE: For packaged mixes, follow directions on package.

SOLUBILITY OF SALT (Sodium Chloride)

Temperature	Amount of salt to saturate 100 g water	
	Grams	%
-4°F (-20°C)	30.9	23.6
68°F (20°C)	36.0	26.4
212°F (100°C)	39.8	28.2

SUBSTITUTION OF INGREDIENTS

For: Substitute:

1 Tbsp flour (15 ml) (used as thickener)
    1 ½ tsp (7 ml) cornstarch, potato starch, rice starch, or arrowroot starch; 1 Tbsp (15 ml) quick cooking tapioca

1 c (250 ml) sifted all-purpose flour
    1 c (250 ml) unsifted all-purpose flour minus 2 Tbsp (30 ml), if spooned into cup

1 c (250 ml) sifted cake flour
    7/8 c (225 ml) sifted all-purpose flour, or 1 c minus 2 Tbsp (225 ml) sifted all-purpose flour

1 c (250 ml) sifted self-rising flour
    1 c (250 ml) sifted all-purpose flour plus 1 ½ tsp (7 ml) baking powder and ½ tsp (2 ml) salt

1 c (250 ml) dark corn syrup
    3/4 c (175 ml) light corn syrup and 1/4 c (50 ml) light molasses

1 c (250 ml) corn syrup
    1 c (250 ml) sugar plus 1/4 c (50 ml) liquid*

1 c (250 ml) light brown sugar
    ½ c (125 ml) dark brown sugar plus ½ c (125 ml) granulated sugar

1 c (250 ml) honey
    1 1/4 c (300 ml) sugar plus 1/4 c (50 ml) liquid*

For: Substitute:

1 ounce (28 g) semisweet chocolate
    ½ ounce (14 g) unsweetened baking chocolate plus 1 Tbsp (15 ml) sugar

1 ounce (28 g) unsweetened chocolate
    3 Tbsp (50 ml) cocoa plus 1 Tbsp (15 ml) fat

1 c (250 ml) coffee cream (20% fat)
    3 Tbsp (50 ml) butter plus about 7/8 c (255 ml) milk

1 c (250 ml) heavy cream (40% fat)
    1/3 c (75 ml) butter plus about 3/4 c (175 ml milk

1c (250 ml) whole milk
    1 c (250 ml) reconstituted nonfat dry milk plus 2 1/2 tsp (12 ml) butter or margarine, or 1/2 c (125 ml) evaporated milk plus 1/2 c (125 ml) water

1 c (250 ml) milk
    3 Tbsp (50 ml) sifted regular nonfat dry milk plus 1 c (250 ml) water, or 1/3 c (75 ml) instant nonfat dry milk plus 1 c (250 ml) minus 1 Tbsp water

1 c (250 ml) buttermilk or sour milk
    1 Tbsp (15 ml) vinegar or lemon juice plus enough sweet milk to make 1 c (250 ml), let stand 5 minutes, or 1 3/4 tsp (8 ml) cream of tartar plus 1 c (250 ml) sweet milk

1 c (250 ml) sour cream
    1 c (250 ml) plain yogurt or 7/8 c (225 ml) sour milk plus 1/3 c (75 ml) butter

1 c (250 ml) buttermilk
    1 c (250 ml) plain yogurt

1 c (250 ml) milk
    1 c (250 ml) buttermilk and omit 1 tsp (5 ml) baking powder, replacing with 1/2 tsp (2 ml) soda and 1 tsp (5 ml) shortening

For:	Substitute:
1 c (250 ml) butter	
	7/8 c (225 ml) oil
1 tsp (5 ml) baking powder	
	1/4 tsp (1 ml) baking soda plus 5/8 tsp (3 ml) cream of tartar; or 1/4 tsp (1 ml) baking soda plus 1/2 cup (125 ml) fully soured milk or buttermilk; or 1/4 tsp (1 ml) baking soda plus 1/2 Tbsp (7 ml) vinegar or lemon juice used with sweet milk to make 1/2 c (125 ml); or 1/4 tsp (1 ml) baking soda plus 1/4 to 1/2 c (50 ml to 125 ml) molasses; or 1/4 tsp (1 ml) baking soda plus 1/2 c (125 ml) yogurt
1 Tbsp (15 ml) active dry yeast	
	1 package active dry yeast, or 1 compressed yeast cake
1 whole egg	
	2 egg yolks, or 3 Tbsp (50 ml) plus 1 tsp thawed frozen egg, or 2 Tbsp (30 ml) and 2 tsp dry whole egg powder plus an equal amount of water
1 egg yolk	
	3 1/2 tsp (17 ml) thawed frozen egg yolk, or 2 tsp (10 ml) dry egg yolk plus 2 tsp (10 ml) water

For:	Substitute:
1 egg white	
	2 Tbsp (30 ml) thawed frozen egg white, or 2 tsp (10 ml) dry egg white plus 2 Tbsp (30 ml) water
1 c (250 ml) bread crumbs	
	3/4 c (175 ml) cracker crumbs
1 c (250 ml) tomato juice	
	1/2 c (125 ml) tomato sauce plus 1/2 c (125 ml) water
1 medium clove garlic	
	1/8 tsp (0.5 ml) garlic powder or 3/4 tsp (3 ml) garlic salt
1 tsp (5 ml) lemon juice	
	1/2 tsp (2 ml) vinegar
1 tsp (5 ml) grated lemon peel	
	1/4 to 1/2 tsp (2 ml) lemon extract
1 small onion	
	1 tsp (5 ml) onion powder or 1 tsp (5 ml) minced dry onion

* Use whatever liquid is called for in the recipe.

NOTE: The amounts of corn syrup and honey are based on the way these products function in recipes and not on the sweetness equivalence with sugar.

TABLE 1.4

Temperature Conversion Table

The numbers in the body of the table give in degrees F the temperature indicated in degrees C at the top and side. To convert 178°C to Fahrenheit scale, find 17 in the column headed degrees C. Proceed in a horizontal line to the column headed 8 which shows 352°F as corresponding to 178°C.

To convert 352°F to Celsius scale, find 352 in the Fahrenheit readings, then in the column headed degrees C, find the number which is on the same horizontal line, i.e., 17. Next, fill in the last number from the heading of the column in which 352 was found, i.e., 8, resulting 178°C which is equivalent to 352°F.

Range:
  -29°C (-20°F) to 309°C (588°F)

Conversion Formulae:
  T°C = 5/9 (T°F-32)
  T°F = 9/5 (T°C+32)

TABLE 1.4 (CONTINUED)

Temperature Conversion Table

Degrees C	0	1	2	3	4	5	6	7	8	9
-2	-4°F	6°F	8°F	-9°F	-11°F	-13°F	-15°F	-17°F	-18°F	-20°F
-1	14°F	12°F	10°F	9°F	7°F	5°F	3°F	1°F	0°F	2°F
-0	32°F	30°F	28°F	27°F	25°F	23°F	21°F	19°F	18°F	16°F
0	32°F	34°F	36°F	37°F	39°F	41°F	43°F	45°F	46°F	48°F
1	50°F	52°F	54°F	55°F	57°F	59°F	61°F	63°F	64°F	66°F
2	68°F	70°F	72°F	73°F	75°F	77°F	79°F	81°F	82°F	84°F
3	86°F	88°F	90°F	91°F	93°F	95°F	97°F	99°F	100°F	102°F
4	104°F	106°F	108°F	109°F	111°F	113°F	115°F	117°F	118°F	120°F
5	122°F	124°F	126°F	127°F	129°F	131°F	133°F	135°F	136°F	138°F
6	140°F	142°F	144°F	145°F	147°F	149°F	151°F	153°F	154°F	156°F
7	158°F	160°F	162°F	163°F	165°F	167°F	169°F	171°F	172°F	174°F
8	176°F	178°F	180°F	181°F	183°F	185°F	187°F	189°F	190°F	192°F
9	194°F	196°F	198°F	199°F	201°F	203°F	205°F	207°F	208°F	210°F
10	212°F	214°F	216°F	217°F	219°F	221°F	223°F	225°F	226°F	228°F
11	230°F	232°F	234°F	235°F	237°F	239°F	241°F	243°F	244°F	246°F
12	248°F	250°F	252°F	253°F	255°F	257°F	259°F	261°F	262°F	264°F
13	266°F	268°F	270°F	271°F	273°F	275°F	277°F	279°F	280°F	282°F
14	284°F	286°F	288°F	289°F	291°F	293°F	295°F	297°F	298°F	300°F
15	302°F	304°F	306°F	307°F	309°F	311°F	313°F	315°F	316°F	318°F
16	320°F	322°F	324°F	325°F	327°F	329°F	331°F	333°F	334°F	336°F
17	338°F	340°F	342°F	343°F	345°F	347°F	349°F	351°F	352°F	354°F
18	356°F	358°F	360°F	361°F	363°F	365°F	367°F	369°F	370°F	372°F
19	374°F	376°F	378°F	379°F	381°F	383°F	385°F	387°F	388°F	390°F
20	392°F	394°F	396°F	397°F	399°F	401°F	403°F	405°F	406°F	408°F
21	410°F	412°F	414°F	415°F	417°F	419°F	421°F	423°F	424°F	426°F
22	428°F	430°F	432°F	433°F	435°F	437°F	439°F	441°F	442°F	444°F
23	446°F	448°F	450°F	451°F	453°F	455°F	457°F	459°F	460°F	462°F
24	464°F	466°F	468°F	469°F	471°F	473°F	475°F	477°F	478°F	480°F
25	482°F	484°F	486°F	487°F	489°F	491°F	493°F	495°F	496°F	498°F
26	500°F	502°F	504°F	505°F	507°F	509°F	511°F	513°F	514°F	516°F
27	518°F	520°F	522°F	523°F	525°F	527°F	529°F	531°F	532°F	534°F
28	536°F	538°F	540°F	541°F	543°F	545°F	547°F	549°F	550°F	552°F
29	554°F	556°F	558°F	559°F	561°F	563°F	565°F	567°F	568°F	570°F
30	572°F	574°F	576°F	577°F	579°F	581°F	583°F	585°F	586°F	588°F

TABLE 1.5

Basic Recipe Proportions

The following table shows proportions of ingredients to one another in certain basic recipes. The amounts of ingredients do not constitute full-size recipes and are not intended for family meal preparation. Weights are given for situations where ingredients are being weighed. Metric conversions can be done on the basis of 1 cup equals 236.6 milliliters and 1 teaspoon equals 5 milliliters.

Product	Flour*	Baking Liquid	Other Fat	Eggs	Sugar	Salt	Powder†	Ingredients
BEVERAGES								
Cocoa and Chocolate		1 c milk 242 g			2 tsp – 1 Tbsp 8.3 – 12.5 g	Few grains		1 Tbsp cocoa 7 g or 1/2 oz chocolate 14.2 g
Coffee		3/4 c water 178 g						1 Tbsp coffee 5.3 g
Instant		3/4 c water 178 g						1 – 2 tsp instant coffee 0.8 – 1.6 g
Freeze dried		3/4 c water 178 g						1/3 tsp 0.5 g
Tea		3/4 c water 178 g						1/2 – 1 tsp tea 0.75 g – 1.5 g
BREADS								
Biscuits	1 c 115 g	1/3 – 1/2 c milk 12.6 – 47.2 g	2 – 4 Tbsp 23.6 – 47.2 g			1/2 tsp 3 g	1 1/4 or 2 tsp 4.5 or 5.8 g	
Griddle cakes (pancakes)	1 c 115 g	3/4 – 7/8 c milk 181.5 – 211.8 g	1 Tbsp 11.8 g	1/2 25 g	0 – 1 Tbsp 0 – 12.5 g	1/2 tsp 3 g	1 1/2 or 2 tsp 5.4 or 5.8 g	
Muffins	1 c 115 g	1/2 c milk 121 g	2 – 3 Tbsp 23.6 – 35.4 g	1/2 25 g	1 – 2 Tbsp 12.5 – 25 g	1/2 tsp 3 g	1 1/4 or 2 tsp 4.5 or 5.8 g	
Popovers	1 c 115 g	1 c milk 242 g	1 – 2 Tbsp 11.8 – 23.6 g			1/4 – 3/4 tsp 1.5 – 4.5 g		
Waffles	1 c 115 g	3/4 – 1 c milk 181.5 – 242 g	1 – 3 Tbsp 11.8 – 35.4 g	1 – 2 50 – 100 g		1/2 tsp 3 g	1 1/4 or 2 tsp 4.5 or 5.8 g	

(continued on next page)

TABLE 1.5 (CONTINUED)
Basic Recipe Proportions

Product	Flour*	Liquid	Fat	Eggs	Sugar	Salt	Baking Powder†	Other Ingredients
Yeast Bread	1 c 115 g	1/3 c milk 80.7 g	0 – 1 Tbsp 0 – 11.8 g		1 tsp – 1 Tbsp 4.2 – 12.5 g	1/4 tsp 1.5 g		1/4 compressed yeast cake 12.8 g or 1/4 small package active dry yeast 1.7 g
CAKE AND PASTRY								
Cake with fat	1 c cake or all-purpose 96 or 115 g	1/4 – 1/2 c milk 60.5 – 121 g	2 – 4 Tbsp 23.6 – 47.2 g	1/2 – 1 25 – 50 g	1/2 – 3/4 c 100 – 150 g	1/8 – 1/3 tsp 0.75 – 1.5 g	1 or 2 tsp 3.6 – 5.8 g	Flavoring
Chiffon	1 c cake 96 g	1/3 c water 79 g	1/4 c (salad oil) 52.5 g	1 150 g	2/3 c 134 g	1/2 tsp 3 g	1 1/4 or 1 1/2 tsp 4.5 – 5.4 g	1/4 tsp cream of tarter 0.8 g Flavoring
Cake without added fat								
Angel food	1 c cake 96 g			1 – 1 1/2 c whites 246-369 g	1 1/4 – 1 1/2 c 250 – 300 g	1/2 tsp 3 g		3/4 – 1 1/2 tsp cream of tarter 2.3 – 4.6 g Flavoring
Sponge	1 c cake 96 g	0 – 3 Tbsp water 0 – 44.4 g		5 – 6 250-300 g	1 c 200 g	1/2 tsp 3 g		0 – 3/4 tsp cream of tarter 0 – 2.3 g Flavoring
Cream Puffs	1 c 115 g	1 c water 237 g	1/2 c 144 g	4 200 g		1/4 tsp 1.5 g		
Doughnuts	1 c 115 g	1/4 c milk 60.5 g	1 tsp 3.9 g	1/2 25 g	1/4 c 50 g	1/4 tsp 1.5 g	1 or 2 tsp 3.6 – 5.8 g	Flavoring
Pastry	1 c 115 g	2 – 3 Tbsp water 29.6 g	4 – 5 1/2 Tbsp 47.2-58.8 g			1/2 tsp 3 g		
EGG DISHES								
Custards		1 c milk 242 g		1 – 1 2/3 50 – 83.5 g	1 1/2 – 3 Tbsp 18.8 – 37.5 g	1/8 tsp 0.75 g		Flavoring
Omlets		1 Tbsp 151.1 g		1 50 g		1/8 tsp 0.75 g		Seasonings
Soufflés	1/4 – 1/3 c							
Entrée	3/4 Tbsp 21.6 – 28.8 g	1 c milk 242 g	3 – 4 Tbsp 35.2 – 47.2 g	3 150 g		1/4 – 1/2 tsp 1.5 – 3 g		Seasonings

(continued on next page)

TABLE 1.5 (CONTINUED)
Basic Recipe Proportions

Product	Flour*	Liquid	Fat	Eggs	Sugar	Salt	Baking Powder†	Other Ingredients
PUDDINGS								
Cornstarch		1 c milk 242 g		0 – 1 0 – 50 g	2 – 3 Tbsp 25 – 37.5 g	1/8 tsp 0.75 g		1 – 1 1/2 Tbsp cornstarch 8 – 12 g Flavoring
Gelatin, Jellies								
Plain		2 c milk, water, fruit juices or other liquid 484 g			2 Tbsp 25 g			1 envelope gelatin 1 Tbsp 7 – 10 g
Fruit or vegetable		3 c liquid 741 g			1/4 c 50 g	Few grains		2 envelopes gelatin 2 Tbsp 14 – 20 g
Bavarian Creams		2 c milk 484 g		4 200 g	1/4 c 50 g			2 envelopes gelatin – 2 Tbsp 14 – 20 g 2 c fruit pulp 484 g 1 c whipping cream – 242 g Flavoring
Tapioca		1 c milk 242 g		1/2 – 1 25-50 g	2 Tbsp 25 g	1/8 tsp 0.75 g		1 1/2 Tbsp quick cooking tapioca 14.25 g Flavoring
Rice (steamed)		1 c milk			1 – 2 Tbsp 12.5 – 25 g	1/8 tsp 0.75 g		2 – 4 Tbsp raw rice 22.8 – 45.5 g Flavoring
Rice (baked)		1 c milk 242 g	1 Tbsp 11.8 g		1 Tbsp 12.5 g	Few grains		1 Tbsp raw rice 11.4 g Flavoring

(continued on next page)

TABLE 1.5 (CONTINUED)
Basic Recipe Proportions

Product	Flour*	Liquid	Fat	Eggs	Sugar	Salt	Baking Powder†	Other Ingredients
SAUCES								
White wine sauce								
Thin	1 Tbsp 7.8 g	1 c milk 242 g	1 Tbsp 11.8 g			¼ tsp 1.5 g		Pepper, if desired
Medium	2 Tbsp 15.6 g	1 c milk 242 g	2 Tbsp 23.6 g			¼ tsp 1.5 g		Pepper, if desired
Thick	3 – 4 Tbsp 23.4 – 31.2 g	1 c milk 242 g	3 – 4 Tbsp **35.4 – 47.2 g**			¼ tsp 1.5 g		Pepper, if desired
Fruit sauce		1 c fruit juice 247 g			1 – 4 Tbsp 25 – 50 g			¼ – 1 Tbsp cornstarch 6 – 8 g Fruit, if desired
SOUPS								
Cream with nonstarchy vegetables	1 Tbsp 7.8 g	1 c milk 242 g	1 Tbsp 11.8 g			¼ tsp 1.5 g		Seasonings, vegetables
Cream with starch vegetables	1½ tsp 3.7 g	1 c milk 242 g	1 Tbsp 11.8 g			¼ tsp 1.5 g		Seasonings, vegetables

* All-purpose flour unless cake flour is specified.

† Use the smaller amount with SAS-phosphate powder and the larger amount with quick acting powder.

TABLE 1.6

Table of Equivalents

ABBREVIATIONS AND SYMBOLS*

Capacity
Cup (c)
Deciliter (d)
Fluid ounce (fl oz)
Gallon (gal)
Liter (l)
Milliliter (ml)
Pint (pt)
Quart (qt)
Peck (pk)
Bushel (bu)
Gill (gi)
Tablespoon (Tbsp)
Teaspoon (tsp)

Time
Hour (hr)
Minute (min)
Second (sec)

Temperature
Degrees Celsius (°C)
Degrees Fahrenheit (°F)

Length
Centimeter (cm)
Foot (ft)
Inch (in)
Meter (m)
Millimeter (mm)
Millimicron (mμ)

Weight
Gram (g)
Kilogram (kg)
Microgram (μg)
Milligram (mg)
Ounce (oz)
Pound (lb)

* Note that abbreviations are used in the singular form regardless of whether the item is singular or plural. For metric units a symbol is used, as g for grams.

WEIGHT AND VOLUME EQUIVALENTS

Common Units of Weight

1 gram	=	0.035 ounces
1 kilogram	=	2.21 pounds
1 ounce	=	28.35 grams
1 pound	=	453.59 grams

Common Units of Volume

1 bushel	=	4 pecks
1 peck	=	8 quarts
1 gallon	=	4 quarts
1 quart	=	2 pints
	=	946.4 milliliters
1 pint	=	2 cups
1 cup	=	16 tablespoons
	=	2 gills
	=	8 fluid ounces
	=	236.6 milliliters
1 Tablespoon	=	3 teaspoons
	=	1/2 fluid ounce
	=	14.8 milliliters
1 teaspoon	=	4.9 milliliters
1 liter	=	1000 milliliters
	=	1.06 quarts

EQUIVALENTS FOR ONE UNIT AND FRACTIONS OF A UNIT

Tablespoon

1 Tbsp = 3 tsp	1/2 Tbsp = 1 1/2 tsp
7/8 Tbsp = 2 1/2 tsp	3/8 Tbsp = 1 1/8 tsp
3/4 Tbsp = 2 1/4 tsp	1/3 Tbsp = 1 tsp
2/3 Tbsp = 2 tsp	1/4 Tbsp = 3/4 tsp
5/8 Tbsp = 1 7/8 tsp	

Cup

1 c = 16 Tbsp	3/8 c = 6 Tbsp
7/8 c = 14 Tbsp	1/3 c = 5 1/3 Tbsp
3/4 c = 12 Tbsp	1/4 c = 4 Tbsp
2/3 c = 10 2/3 Tbsp	1/8 c = 2 Tbsp
5/8 c = 10 Tbsp	1/16 c = 1 Tbsp
1/2 c = 8 Tbsp	

Pint

1 pt = 2 c	3/8 pt = 3/4 c
7/8 pt = 1 3/4 c	1/3 pt = 2/3 c
3/4 pt = 1 1/2 c	1/4 pt = 1/2 c
2/3 pt = 1 1/3 c	1/8 pt = 1/4 c
5/8 pt = 1 1/4 c	1/16 pt = 2 Tbsp
1/2 pt = 1 c	

Quart

1 qt = 2 pt	3/8 qt = 1 1/2 c
7/8 qt = 3 1/2 c	1/3 qt = 1 1/3 c
3/4 qt = 3 c	1/4 qt = 1 c
2/3 qt = 2 2/3 c	1/8 qt = 1/2 c
5/8 qt = 2 1/2 c	1/16 qt = 1/4 c
1/2 qt = 1 pt	

Gallon

1 gal = 4 qt	3/8 gal = 3 pt
7/8 gal = 3 1/2 qt	1/3 gal = 5 1/3 c
3/4 gal = 3 qt	1/4 gal = 1 qt
2/3 gal = 10 2/3 c	1/8 gal = 1 pt
5/8 gal = 5 pt	1/16 gal = 1 c
1/2 gal = 2 qt	

Pound

1 lb = 16 oz	3/8 lb = 6 oz
7/8 lb = 14 oz	1/3 lb = 5 1/3 oz
3/4 lb = 12 oz	1/4 lb = 4 oz
2/3 lb = 10 2/3 oz	1/8 lb = 2 oz
5/8 lb = 10 oz	1/16 lb = 1 oz
1/2 lb = 8 oz	

TABLE 1.7

Cooking Conversions

U.S. to Metric

Capacity		Weight	
1/5 teaspoon	1 milliliter	1 ounce	28 grams
1 teaspoon	5 milliliters	1 pound	454 grams
1 tablespoon	15 milliliters		
1 fluid ounce	30 milliliters		
1/5 cup	47 milliliters		
1 cup	237 milliliters		
2 cups (1 pint)	473 milliliters		
4 cups (1 quart)	.95 liter		
4 quarts (1 gallon)	3.8 liters		

Metric to U.S.

Capacity		Weight	
1 milliliter	1/5 teaspoon	1 gram	.035 ounce
5 milliliters	1 teaspoon	100 grams	3.5 ounces
15 milliliters	1 tablespoon	500 grams	1.10 pounds
100 milliliters	3.4 fluid ounces	1 kilogram	2.205 pounds
240 milliliters	1 cup		35 ounces
1 liter	34 fluid ounces		
	4.2 cups		
	2.1 pints		
	1.06 quarts		
	0.26 gallon		

TABLE 1.8

Conversion to Metric Units

Comparison of Avoirdupois and Metric Units of Weight

1 oz	=	0.06 lb	=	28.35 g	=	(30 g)*	1 g =	0.035 oz
2 oz	=	0.12 lb	=	56.70 g			2 g =	0.07 oz
3 oz	=	0.19 lb	=	85.05 g			3 g =	0.11 oz
4 oz	=	0.25 lb	=	113.40 g	=	(125 g)	4 g =	0.14 oz
5 oz	=	0.31 lb	=	141.75 g			5 g =	0.18 oz
6 oz	=	0.38 lb	=	170.10 g			6 g =	0.21 oz
7 oz	=	0.44 lb	=	198.45 g			7 g =	0.25 oz
8 oz	=	0.50 lb	=	226.80 g	=	(250 g)	8 g =	0.28 oz
9 oz	=	0.56 lb	=	255.15 g			9 g =	0.32 oz
10 oz	=	0.62 lb	=	283.50 g			10 g =	0.35 oz
11 oz	=	0.69 lb	=	311.85 g			11 g =	0.39 oz
12 oz	=	0.75 lb	=	340.20 g	=	(350 g)	12 g =	0.42 oz
13 oz	=	0.81 lb	=	368.55 g			13 g =	0.46 oz
14 oz	=	0.88 lb	=	369.90 g			14 g =	0.49 oz
15 oz	=	0.94 lb	=	425.25 g			15 g =	0.53 oz
16 oz	=	1.00 lb	=	453.59 g	=	(500 g)	16 g =	0.56 oz

* Figures in parentheses indicate common use.

Comparison of U.S. and Metric Units of Liquid Measure

1 fl oz	=	29.573 ml	1 qt	=	0.946 l	1 gal	=	3.785 l
2 fl oz	=	59.15 ml	2 qt	=	1.89 l	2 gal	=	7.57 l
3 fl oz	=	88.72 ml	3 qt	=	2.84 l	3 gal	=	11.36 l
4 fl oz	=	118.30 ml	4 qt	=	3.79 l	4 gal	=	15.14 l
5 fl oz	=	147.87 ml	5 qt	=	4.73 l	5 gal	=	18.93 l
6 fl oz	=	177.44 ml	6 qt	=	5.68 l	6 gal	=	22.71 l
7 fl oz	=	207.02 ml	7 qt	=	6.62 l	7 gal	=	26.50 l
8 fl oz	=	236.59 ml	8 qt	=	7.57 l	8 gal	=	30.28 l
9 fl oz	=	266.16 ml	9 qt	=	8.52 l	9 gal	=	34.07 l
10 fl oz	=	295.73 ml	10 qt	=	9.46 l	10 gal	=	37.85 l
1 ml	=	0.034 fl oz	1 l	=	1.056 qt	1 l	=	0.264 gal
2 ml	=	0.07 fl oz	2 l	=	2.11 qt	2 l	=	0.53 gal
3 ml	=	0.10 fl oz	3 l	=	3.17 qt	3 l	=	0.79 gal
4 ml	=	0.14 fl oz	4 l	=	4.23 qt	4 l	=	1.06 gal
5 ml	=	0.17 fl oz	5 l	=	5.28 qt	5 l	=	1.32 gal
6 ml	=	0.20 fl oz	6 l	=	6.34 qt	6 l	=	1.59 gal
7 ml	=	0.24 fl oz	7 l	=	7.40 qt	7 l	=	1.85 gal
8 ml	=	0.27 fl oz	8 l	=	8.45 qt	8 l	=	2.11 gal
9 ml	=	0.30 fl oz	9 l	=	9.51 qt	9 l	=	2.38 gal
10 ml	=	0.34 fl oz	10 l	=	10.57 qt	10 l	=	2.64 gal

TABLE 1.9

Hydrogen Ion Concentration and pH of Some Common Foods

Common Indicator Changes*	Hydrogen Ion Concentration	pH	Average Values for Common Foods
R at ph 1.2	1.0×10^{-2}	2.0	Limes
	8.0×10^{-3}	2.1	
	6.3×10^{-3}	2.2	Lemons
	5.0×10^{-3}	2.3	
	4.0×10^{-3}	2.4	
Thymol blue	3.2×10^{-3}	2.5	
	2.5×10^{-3}	2.6	
	2.0×10^{-3}	2.7	
	1.6×10^{-3}	2.8	
	1.3×10^{-3}	2.9	Vinegar, plums
	1.0×10^{-3}	3.0	Gooseberries
		3.1	Prunes, apples grapefruit (3.0 to 3.3)
		3.2	Rhubarb, dill pickles
		3.3	Apricots, blackberries
		3.4	Strawberries, lowest acidity for jelly
		3.5	Peaches
		3.6	Raspberries, sauerkraut
		3.7	Blueberries, oranges (3.1 to 4.1)
Bromphenol blue		3.8	Sweet cherries
		3.9	Pears
	1.0×10^{-4}	4.0	Acid fondant, acidophilus milk
		4.1	
		4.2	Tomatoes (4.0 to 4.6)
		4.3	
		4.4	Lowest acidity for processing at 100°C
Bromcresol green		4.5	Buttermilk
		4.6	Bananas, egg albumin, figs, isoelectric point for casin, pimentos
		4.7	
		4.8	
		4.9	
	1.0×10^{-5}	5.0	Pumpkins, carrots
		5.1	Cucumbers
		5.2	Turnips, cabbage, squash
Methyl red		5.3	Parsnips, beets, green peppers, string beans
Chlorophenol red		5.4	Sweet potatoes, bread
		5.5	Spinach
		5.6	Asparagus, cauliflower
		5.7	Red kidney beans, lima beans
		5.8	Meat, ripened, poultry, succotash
		5.9	
Bromcresol purple	1.0×10^{-6}	6.0	Tuna
		6.1	Potatoes
		6.2	Peas
		6.3	Corn, oysters, dates
		6.4	Egg yolks
		6.5	
		6.6	
Litmus		6.7	Milk (6.5 to 6.7)
		6.8	
		6.9	Shrimp, wet packed
Bromthymol blue	1.0×10^{-7}	7.0	Meat, unripened
		7.1	Lye, hominy
		7.2	
		7.3	
		7.4	
		7.5	
Phenol red		7.6	
		7.7	
		7.8	
		7.9	
	1.0×10^{-8}	8.0	Egg white
		8.1	
		8.2	
		8.3	
		8.4	

* COMMON INDICATOR CHANGES:
B = Blue
P = Purple
R = Red
Y = Yellow

TABLE 1.10

Temperature of Food for Control of Bacteria

°C	°F	
121	250	121°C to 116°C – Canning temperature for low-acid vegetables, meat, fish and poultry in pressure canner.
116	240	
		106°C to 102°C – Jam-Jelly stage
100	212	**Boiling Point**
		100°C – Temperature of boiling water at sea level. Temperature to can fruits, jams, jellies, preserves, low-acid tomatoes, and pickles in a water-bath canner.
85		
80	180	100°C to 85°C – Simmering
		80°C – Scalding
		Cooking temperatures destroy most bacteria. The time required to kill bacteria is less as temperature increases.
		83°C to 60°C – Pasteurizing
74	165	Warming temperatures prevent bacterial growth, but allow the survival of some bacteria.
60	140	Some bacterial growth may occur at this temperature. Many bacteria survive.
	125	
		DANGER ZONE: Temperatures in this zone allow rapid growth of bacteria and production of toxins by some bacteria. (Do not keep foods in this temperature zone for more than 2-3 hours.)
35	95	35°C – Lukewarm
30	86	26°C to 32°C – yeast dough rises
	60	15°C to 60°C – rennin, an enzyme, is active
10	50	60°C to 4°C – These temperatures allow the growth of bacteria and mold that cause foodborne illness.
4	40	
2	36	4 °C – Optimum refrigerator storage. Temperature for thawing frozen foods. Cold temperatures permit slow growth of some bacteria that cause spoilage.
0	32	**Freezing Point**
		Freezing temperatures stop growth of bacteria, but may not kill them.
-18	0	
		-18°C to -29°C – Optimum freezer storage.
-29	-20	

Adapted from *You Can Prevent Food Poisoning.*
Pacific Northwest Publication No. 250.
Coop. Ext. WSU, Pullman, WA, 1989.

TABLE 1.11
Thickening and Gelling Agents

Thickening Agent	Uses and Quantity Required	Precautions in Mixing	Effect of Temperature	Other Factors that Affect Thickening	Characteristics of the Gel
AGAR	**Salad and desert jellies:** 4 to 6 g (about 2 tsp) per pt of liquid (500 ml).	Soak in 3 to 6 times the weight of cold liquid, then dissolve by bringing to a boil.	Gel forms on cooling to 40°C to 45°C (104°F to 113°F); softens at 80°C to 85°C (176°F to 185°F).	Gel strength not easily destroyed by heat or acid.	Gel rigid, short, crumbly; transparent; may have weedy odor if sample not highly purified.
EGG	**Custard puddings and sauces:** 2 to 3 medium eggs per pt (500 ml) of milk or 4 to 6 yolks.	Blend well with sugar and milk. Coagulate by slow heating.	Overcooking causes syneresis (weeping) and curdling.	Sugar and dilution raise coagulation temperature; acid lowers it. Use of water instead of milk results in a flocculent precipitate rather than a gel.	Baked custard forms firm, continuous clot. Stirred custard is soft, thickened but not set.
FLOUR	**Thin soup or sauce:** 2 Tbsp (16 g) per pt (500 ml) of liquid. **Medium sauces:** 4 to 5 Tbsp (32 to 40 g) per pt (500 ml) of liquid. **Soufflés, molded pastes:** 1/2 c per pt of liquid (125 ml/500 ml).	Disperse in cold liquid or in fat or mix with sugar before adding hot liquid. Stir while cooking.	Heat to 90°C (194°F) or above to obtain maximum thickening. Viscosity increases on cooling.	Heating with acid causes thinning. High sugar concentrations retard gelatinization and reduce thickening power.	Opaque paste.
WAXY RICE FLOUR (mochiko, sweet rice flour)	**Frozen sauces and gravies:** to prevent curdling and liquid separation 4 to 5 Tbsp per pt (32 to 40 g/500 ml) liquid. Also prevents gelation of thickening canned products.	Same as flour, but waxy rice is less likely to lump.	Maximum thickening at 70°C to 80°C (158°F to 176°F); little difference between hot and cold viscosity.	Heating with acid and homogenization cause thinning.	Does not gel. Forms short opaque paste.

(continued on next page)

TABLE 1.11 (CONTINUED)
Thickening and Gelling Agents

Thickening Agent	Uses and Quantity Required	Precautions in Mixing	Effect of Temperature	Other Factors that Affect Thickening	Characteristics of the Gel
GELATIN	**Molded deserts and salads:** 7 to 12 g (about 1 Tbsp or 1 envelope) per pt (500 ml) of liquid according to grade.	Soak in 2 to 6 times the weight of cold liquid, then dissolve by heating to 40°C (104°F) or by adding hot liquid, and then add sugar.	Gel forms after a few hours of chilling. Softens at 26.5°C (80°F) and higher.	Heating with acid causes reduction of gel strength. Raw pineapple, kiwi, and figs prevent setting because of enzyme action.	Gel firm but springy and quivery; transparent in appearance.
GUM TRAGACANTH	**Salad dressings, sauces:** 2 to 3 g per pt (500 ml) of liquid give thin paste; 6 to 8 g per pt (500 ml) give thick gel.	Dissolve either in hot or cold water (dissolves much more rapidly in hot).	Little change in viscosity over a wide temperature range.	Acid, alkali or salt plus heat causes thinning.	Gel thick and mucilaginous but not rigid even at high concentrations.
IRISH MOSS	**Puddings, sauces:** 4 to 6 g per pt (500 ml) liquid give a thick paste; 15 to 25 g per pt (500 ml) of liquid give a stiff gel.	Soak in cold water, then heat to 60°C (140°F) or above to dissolve.	Gel melts at 27°C to 41°C (81°F to 106°F) depending on concentration.	Acid plus heat causes thinning. Heat alone has no effect.	Gel thick and mucilaginous; becomes rigid only at high concentrations; may have weedy odor if not highly purified.
CORNSTARCH AND RICE STARCH	1 Tbsp cornstarch or rice starch = 2 Tbsp flour.	Same as flour.	Same as flour.	Same as flour.	Pastes more translucent than flour paste.
POTATO STARCH and ARROWROOT STARCH	1 Tbsp potato or arrowroot starch = 2 Tbsp flour (See *Effect of Temperature column*). Suitable for starch-egg mixtures or fruit sauces where higher temperatures are not desired.	Same as flour.	Reached maximum thickening at 70°C to 80°C (158°F to 176°F); higher temperature or further heating causes very marked thinning.	Same as flour, thinning also brought about by excessive stirring.	Paste very transparent.

(continued on next page)

Thickening Agent	Uses and Quantity Required	Precautions in Mixing	Effect of Temperature	Other Factors that Affect Thickening	Characteristics of the Gel
WAXY CEREAL STARCHES	Prevent gelatin and syneresis of canned products during storage—may be used in combination with flour.	Same as flour.	Similar to waxy rice flour.	Thinned by heating with acid.	Waxy starches give somewhat ropy translucent pastes.
TAPIOCA, PEARL	About twice as much as quick-cooking tapioca.	Soak several hours before cooking.	Cook until tapioca is transparent.	Same as quick-cooking tapioca.	Same as quick-cooking tapioca.
TAPIOCA, QUICK-COOKING	**Puddings:** 3 Tbsp per pt (500 ml) of liquid. **Fruit pie fillings:** 1 1/2 to 3 Tbsp for 8- or 9-inch pie. **Soup:** 1 1/2 to 3 Tbsp per qt (100 ml) of liquid.	Mix in cold or hot liquid; no soaking necessary.	Bring only to a boil. Mixture thickens as tapioca particles swell and become transparent. It continues to thicken while cooling.	Stir while cooking. Over-stirring while cooling tends to disrupt tapioca particles, resulting in a sticky, gelatinous mixture.	Pastes transparent, nonhomogenous (particles remain distinct). Mixture thickens as it cools. Especially satisfactory for fruit pie fillings.

FOODBORNE ILLNESSES AND THEIR PREVENTION

Foodborne disease is estimated to cause many illnesses, hospitalizations, and even deaths annually in the United States. Illnesses occur from improper food handling practices in consumer homes and foodservice establishments, both commercial and institutional. Some population groups such as immune-compromised individuals, older adults, infants and children, and pregnant women are considered to be more highly susceptible to foodborne illness than other individuals.

The majority of cases of foodborne illness are caused by bacteria and viruses. Parasites and fungi are less common causes. Bacterial foodborne illnesses can be classified as foodborne infections, foodborne intoxications, or foodborne toxin-mediated infections. *The Bad Bug Book* on the FDA website (www.cfsan.fda.gov~mow/intro/html) lists specific information about a variety of foodborne pathogens.

Foodborne infections occur when humans consume a food or beverage containing significant numbers of bacterial cells, and the bacteria then reproduces inside the human body. Most foodborne infections result in symptoms of abdominal pain, headache, fever, nausea, and diarrhea. Incubation periods vary according to the type of bacteria involved, and duration can range from 1 day to a week. Some of the most common bacteria causing foodborne infections and the foods in which they occur are shown below.

TABLE 1.12

Examples of Bacteria Causing Foodborne Infections and Associated Foods

Campylobacter jejuni	Raw chicken, raw milk, and untreated drinking water.
E. Coli O157:H7	Raw and undercooked beef, alfalfa sprouts, unpasteurized fruit juice, dry-cured salami, lettuce, game meta, cheese curds, and raw milk.
Listeria monocytogenes	Raw milk, soft cheeses, ice cream, raw vegetables, fermented raw-meat sausages, raw and cooked poultry, raw meats, and raw and smoked fish.
Salmonella	Raw meats, poultry, eggs, milk and dairy products, fish, shrimp, frog legs, meat, coconut, sauces and salad dressings, cake mixes, cream-filled desserts, dried gelatin, peanut butter, cocoa, and chocolate.
Vibrio vulnificus	Oysters, clams, and crab.

Foodborne intoxications are caused when food is improperly handled, and bacteria present in the food grow and produce toxins. When humans ingest the food containing the toxin, they become ill, usually within 6 to 24 hours. Foodborne intoxications often cause symptoms of nausea, vomiting, diarrhea, but not fever. In some cases, such as with *Clostridium botulinum,* death may result. Some of the most common bacteria causing foodborne intoxications are shown below.

TABLE 1.13

Examples of Bacteria Causing Foodborne Intoxications and Associated Foods

Bacillus cereus	Meats, milk, vegetables, fish, rice, starchy foods (potato, pasta, and cheese products), sauces, puddings, soups, casseroles, pastries, and salads.
Clostridium botulinum	Improperly processed, canned low-acid foods (vegetables, fish, poultry and meat).
Staphylococcus aureus	Meat and meat products, poultry and egg products, salads (protein, potato, and pasta), cream-filled pastries and cream pies, sandwich fillings, and milk and dairy products.

Foodborne toxin-mediated infections are caused when bacteria are ingested with food, but once inside the human body grow and produce a toxin that causes illness. *Clostridium perfringens,* which is found in meat, meat products and gravy, produces this kind of illness. *Bacillus cereus* and *E. Coli O157:H7,* which cause foodborne infections, can also cause foodborne toxin-mediated infections. In children, *E. Coli O157:H7* can cause hemolytic uremic syndrome and death.

Viruses carried in food are also a common cause of foodborne illness. Unlike bacteria, viruses can only reproduce in living cells; they do not reproduce in food. Viruses are usually spread to food because of the poor personal hygiene of a food handler who is ill. They may contaminate any type of food, not just those foods that are considered potentially hazardous. Foods commonly implicated in viral foodborne illness include raw and ready-to-eat foods that are not cooked prior to consumption such as salads, raw fruits and vegetables, and sandwiches. Water and ice can also be carriers for viruses that cause illness. Viruses that commonly cause short-term illness (1 to 3 days) are Norwalk and Norwalk-like virus and Rotavirus. Another virus, Hepatitis A, can cause longer term illness ranging from 1 to 2 weeks or several months.

FDA Food Code

The *FDA Food Code,* a comprehensive guide on food safety and sanitation standards and food handling practices, is often used as the basis for state and local regulations and ordinances intended to promote public food safety. The *Food Code* is published every five years with periodic updates during the interim periods. The preface of the *Food Code* states that five major causes of foodborne illness in foodservice establishments are improper holding temperatures, inadequate cooking, contaminated equipment, food from unsafe sources, and poor personal hygiene of food handlers. Some key provisions of the current food code (www.cfsan. fda.gov/~dms/foodcode.html) that apply to managers and staff of foodservice operations are as follows:

- The person in charge of a food establishment must demonstrate knowledge of prevention of foodborne disease, preparation and storage of potentially hazardous foods, cleaning/sanitizing procedures, critical control points, major food allergens, and understanding of the *Food Code.*

- Employers must screen employees for illness from common food pathogens (E. Coli, Salmonella, Hepatitis A, Shigella, and Norovirus) and exclude or restrict those who are ill.

- Employees must wash hands with a 20-second hand-washing at appropriate times.

- Employees must not touch ready-to-eat food with bare hands (except raw fruits and vegetables while washing). Foods should be handled with suitable utensils such as deli tissue, spatula, tongs, single-use gloves, or dispensing equipment.

Some key provisions of the *Food Code* that apply to foods, food handling, and food storage are:

- Potentially hazardous foods include animal food raw or heat-treated (i.e., meat, poultry, seafood, eggs, and dairy) and plant foods that are heat treated. Raw seed sprouts, cut melons, and garlic-in-oil mixtures and other foods with specific combinations of high Aw (> 0.88) and low acidity (pH > 4.2) may also be considered potentially hazardous.

- Only pasteurized fruit juices and eggs should be served to highly susceptible populations.

- Shell eggs are to be stored and transported at temperatures no greater than 45°F.

- Three acceptable ways of thawing frozen foods are: place in a refrigerator, completely submerge under running water at 70°F or less, or thaw as part of the cooking process.

- Hot foods should be held at 135°F or above and cold foods held at 41°F or below.

- Leftover hot cooked foods must be cooled from 135° to 70°F within 2 hours; from 70° to 41°F within 4 hours. Hot cooked foods can be cooled by placing in shallow pans and using rapid cooling methods (ice bath or blast chiller).

- Wet wiping cloths used for counter tops should be free of food debris and visible soil and should be stored in a solution with chemical sanitizer at proper concentration.

Hazard Analysis Critical Control Point (HACCP)

The *Food Code* also emphasizes use of Hazard Analysis Critical Control Point (HACCP) methods to prevent or reduce risk of potential food safety problems. The seven basic steps in HACCP are:

- Identify hazards (biological, chemical, and/or physical) and assess level of risk.

- Identify the critical control points (CCP)—a point or procedure where loss of control may results in an unacceptable health risk.

- Establish critical limits for preventive measures (time and end-point cooking temperatures)

- Establish procedures to monitor CCPs.

- Establish corrective actions to be taken when a critical limit has been exceeded.

- Establish effective record-keeping systems to document HAACP system.

- Establish procedures to verify that the system is working.

Several segments of the food industry are required to use HACCP programs in food processing. FDA regulates HACCP for the seafood industry and for companies that process fruit and vegetable juices and juice products. The Food Safety and Inspection Service (FSIS) of the U.S. Department of Agriculture regulates HACCP for meat and poultry processors. FDA encourages foodservice establishments to adopt a HACCP system and has published a guide for this purpose. Beginning in the 2005-2006 year, all schools participating in the National School Lunch Program and/or National School Breakfast Program were required by federal law to develop a food safety program that complies with HACCP principles.

Consumer Food Safety

While the *Food Code* serves as a guide for foodservice establishments outside the home, other food safety education materials have been developed to educate consumers about food safety. The Partnership for Food Safety Education, which includes the U.S. Food and Drug Administration, U.S. Department of Agriculture, Centers for Disease Control, National Restaurant Association Educational Foundation, International Food Information Council, Institute of Food Technologists, American Dietetic Association, and 19 other professional and trade associations developed the Fight Bac campaign in 1997. Fight Bac (www.fightbac.org) divides consumer food safety education into four main areas:

- Clean: focuses on personal hygiene and cleaning of kitchen counters, utensils, equipment, and raw fruits and vegetables.

- Separate: advises consumers to avoid cross-contamination between potentially hazardous foods, hands, and equipment.

- Chill: emphasizes chilling foods promptly and properly.

- Cook: focuses on cooking foods and holding foods at appropriate temperatures.

Source: Partnership for Food Safety Information

Glossaries

FOOD SCIENCE TERMS

Acid Substances that furnish hydrogen ions.

Acidulate To make sour or acidic by adding vinegar, lemon juice, or cream of tartar.

Alkali Substances that form hydroxyl ions which can neutralize an acid.

Anthocyanins Pigments (classified as flavonoids) which are responsible for the red/blue colors found in certain vegetables and fruits. They are red in the presence of acid; blue-green in the presence of alkali; water soluble.

Anthoxanthins Pigments (classified as flavonoids) that are white in acid, but turn yellow in the presence of alkali; water soluble.

Antioxidant A substance that inhibits the chemical reaction that causes oxidation.

Ascorbic acid Vitamin C. A vitamin that can be used to inhibit enzymatic browning of light-colored fruit such as apples, pears and bananas; an antioxidant.

Boiling point The temperature at which the atmospheric pressure just exceeds the vapor pressure of a liquid.

Browning reactions Darkening of foods caused by enzymes (light-colored cut fruit or vegetables) or reactions of protein and sugar (baked goods).

Calorie See kilocalorie.

Carotenoids Yellow-to-red pigments found in fruits and vegetables; fat soluble.

Cellulose A polysaccharide that provides structural rigidity in cell walls of plants, fruits, and vegetables. It can be softened by cooking but is not digested by humans.

Chlorophyll The green pigment in vegetables which turns olive-green in the presence of acid and/or heat for a period of time.

Coagulation The change from a fluid state to a thickened curd or clot due to denaturation of protein.

Colloidal dispersion A dispersion of very tiny particles of one substance in another substance; the particles are too large to form a true solution and too small to form a coarse suspension. An example is gelatin dissolved in water.

Controlled atmosphere A long-term storage technique where respiration of products such as apples is retarded by rigidly controlling temperature and humidity and adding carbon dioxide at a controlled level.

Crystallization The formation of crystals resulting from chemical elements or compounds solidifying in an orderly structure.

Dehydration	A method of food preservation where water is removed from food by drying in air, in a vacuum, in inert gas, or by direct application of heat.
Denaturation	A change in the physical structure of protein molecules that alters their characteristics. Heat, for example, denatures the protein in an egg.
Emulsify	To suspend small drops of an immiscible liquid in another, forming an emulsion such as mayonnaise.
Enzymes	Proteins that catalyze chemical reactions that change the color, texture, and flavor of foods.
Enzymatic browning	Discoloration of foods, such as cut fruit, due to oxidation catalyzed by enzymes.
Foam	A type of colloidal dispersion in which bubbles of gas are surrounded by liquid.
Food Standards	Specifications for certain foods such as identity, quality, fill of container and enrichment level which are regulated by governmental agencies.
Freeze drying	A method of preserving food by first freezing and then drying under a vacuum. The freeze-dried product may be stored without refrigeration and is much lighter in weight than frozen products.
Freezing	A method of preserving food by chilling it rapidly at a low temperature.
Gel	A liquid-in-solid colloidal system that does not flow. It can be formed by gelatin, pectin, starch, soured milk, and egg.
Gelatinization	The absorption of liquid by starch, which results in swelling of the granules and thickening of the liquid. Heat is required to accomplish this change.
Gluten	The elastic substance formed from manipulating the viscous and insoluble proteins of wheat flour with some liquid.
Homogenize	To break up into tiny particles of the same size, such as fat in milk.
Hydration	The process of absorbing water.
Hydrogenation	The process of adding hydrogen ions to unsaturated (liquid) fat to form a solid or semi-solid fat such as margarine.
Hydrolysis	The process of splitting compounds into simpler components by adding a molecule of water, a reaction that can be facilitated by acid, heat, enzymes, or light.
Irradiation	A process in which foods can be exposed to ionizing radiation to lengthen storage life. Depending on dose, irradiation can kill insects, inhibit sprouting, and kill microorganisms.
Kilocalorie	The amount of heat required to raise the temperature of 1 kilogram of water (1000 g) 1 degree Celsius.
Kosher	Food produced in accordance with Jewish rabbinical law.
Leavening agent	A substance that increases the volume of a baked product (air, steam, and microbiological

	or chemical agents capable of producing carbon dioxide).
Osmosis	Movement of water through a semipermeable membrane from an area of low concentration of solute to an area of higher concentration to equalize the osmotic pressure created by differences in concentration.
Pasteurization	The process of destroying harmful microorganisms by heating food (such as milk or fruit juice) to a specific temperature and holding for a specific length of time.
pH	A measure of acidity/alkalinity based on the hydrogen ion concentration. On the scale of 1 to 14, 1 is most acidic and 14 is most alkaline. Neutral is a pH of 7.
Pickling	The process of preserving food by using acid and salt to inhibit microbial growth.
Proteolytic enzymes	Enzymes capable of effecting a chemical reaction that splits proteins into simpler components. Examples are papain (from papaya) and bromelin (from pineapple).
Rancidity	Flavor and odor deterioration due to either hydrolysis or oxidation of fats.
Rennet	A substance from calves' stomachs that contains rennin, an enzyme used in cheese making. Now made by factory fermentation.
Retrogradation	Formation of crystalline areas in gelatinized starch mixtures (such as bread products and starch pastes) as they age.
Smoke point	The temperature at which a fat begins to smoke and emit irritating vapors.
Sol	A fluid colloidal system in which the continuous phase is liquid.
Solution	A uniform liquid blend containing a solvent (liquid) and a solute (such as salt or sugar) dissolved in the liquid.
Specific gravity	The weight of a specific volume of a substance in relation to the weight of an equal volume of water. At 40° Celsius, 1 milliliter of water weighs 1 gram.
Specific heat	The heat or thermal capacity of a substance in relation to that of water.
Sterilize	To destroy microorganisms by using heat or irradiation.
Supersaturated solution	A solution which contains more solute than an unheated liquid could hold at a given temperature. It is unstable and prone to crystallization.
Suspension	A substance suspended in a liquid (such as starch in cold water) which will remain combined only if stirred.
Syneresis	Drainage of liquid from a gel when cut or disturbed.
Viscosity	The degree to which a fluid flows readily or resists flow.

COOKING TERMS

Bake	To cook in an oven or oven-type appliance. When applied to meats in uncovered containers, it is generally called roasting.
Barbecue	To roast slowly on a rack or spit over coals or under free flame or an electric heat unit. The food is often brushed with a sauce during cooking.
Baste	To moisten meat or other foods during cooking to add flavor and to prevent drying of the surface. The liquid is usually melted fat, meat drippings, fruit juice, sauce, or water.
Batter	A mixture made of flour, liquid, and other ingredients that can be stirred or poured.
Beat	To apply rapid, regular motions using a wire whisk, spoon, hand beater or mixer to a mixture.
Blanch	To briefly heat vegetables or fruit in steam or boiling water to loosen or remove skins or to inactivate enzymes prior to freezing or drying.
Blend	To mix two or more ingredients thoroughly.
Boil	To cook in water that is at boiling temperature. Bubbles will rise continually and break on the surface.
Braise – *slow cooked chicken*	To cook meat or poultry over low heat in a covered utensil in a small amount of liquid or in steam.
Bread	To coat a food with bread or cracker crumbs or other food. The surface may first be coated with beaten egg or other liquid.
Brine	To soak pickled, fermented, or cured foods in a strong salt solution to add flavor and inhibit microbial growth.
Broil	To cook by direct heat on a rack or spit.
Brown	To cook food with moderate or high heat until it browns in color.
Candy	To cook fruit or vegetables in sugar or syrup.
Caramelize	To heat sugar or foods containing sugar until a brown color and characteristic flavor develop.
Chop	To cut into pieces with a knife or other sharp tool, blender, or food processor.
Coat	To spread food with or dip it into a substance such as flour or a sauce until it is covered.
Combine	To stir two or more ingredients together to form a mixture of uniform consistency.
Cream	To mix one or more foods (usually fat and sugar) with a spoon or mixer until soft and smooth; air is incorporated.
Cube	To cut food into small squares; to cut the surface of meat in a checkered pattern to increase tenderness.

Curdle	To form clots in a smooth liquid (such as milk) by precipitating the protein with heat, acid, or enzymes.
Cure	To preserve beef or poultry with a combination of salt or brine, sodium nitrate and/or nitrite, seasonings, and sometimes smoking. Cured products such as corned beef and ham develop a distinctive pink/red color.
Cut	To divide food materials with a knife or scissors.
Cut In	To distribute solid fat throughout dry ingredients using two knives or a pastry blender until flour-coated fat particles are the desired size.
Deep fry	To cook food in enough hot fat for submersion or flotation.
Dice	To cut into small cubes.
Dilute	To make less strong by adding liquid.
Dissolve	To make a solution such as sugar in water; to melt or liquefy.
Dough	A mixture of flour and water (and usually other ingredients) that is stiff enough to be rolled or kneaded.
Dredge	To cover or coat with flour or other fine substances such as bread crumbs or corn meal.
Dust	To sprinkle lightly with flour or sugar.
Flake	To break into small pieces, usually with a fork.
Fold	To combine one ingredient with another by gently turning the mixture with a spoon or spatula to minimize loss of air. Two motions are used: cutting vertically through the mixture; sliding across the bottom of the bowl and up the other side.
Fricassee	To cook cut pieces of food such as fowl, rabbit, or veal by braising in a sauce.
Fry	To cook in fat. A small amount of fat is used for pan-frying or sauteing; deep-fat fried foods are submerged in fat.
Garnish	To add decorative touches (usually edible) to food that is served.
Glaze	To coat with a glossy mixture that enhances both flavor and appearance of food (such as meats, vegetables, and desserts).
Grate	To produce particles of a specific size by rubbing food (such as carrots or cheese) on a grater (or chopping in a blender or food processor).
Grease	To rub the surface of a pan or dish with fat to prevent food from sticking.
Grill	To cook food on a rack with direct heat.
Grind	To reduce food to small particles by cutting or crushing mechanically in a grinder, blender, or food processor.
Julienne	To cut meat, fruit, or vegetables into long, thin strips.
Knead	To work dough by hand or mechanically to develop the gluten needed for structure of

baked products.

Lard To place strips of fat on top or into gashes in the side of uncooked lean meat or fish to give flavor and prevent drying.

Lukewarm A temperature of about 95°F (35°C). Lukewarm liquids and foods feel neither hot nor cold when in contact with the inside of the wrist.

Macerate To soak foods in a liquid to soften them and to absorb flavor.

Marinate To let food stand in a liquid, which is usually a mixture of oil, lemon juice or vinegar and seasonings.

Meringue A foam of beaten egg whites and sugar that is baked. A soft meringue may be baked on a one-crust pie; a baked hard meringue is used as a shell for berries or other dessert fillings.

Mince To chop or cut into very small pieces.

Mix To combine two or more ingredients into a uniform mixture.

Pan-broil To cook uncovered over high heat on a hot surface (usually a fry pan), pouring off accumulating fat.

Pan-fry To cook over high heat in a small amount of fat.

Panning Cooking vegetables in their own juices and a small amount of fat in a tightly covered pan.

Parboil To boil until partially cooked. Cooking is usually completed by another method.

Pare To cut off the outside covering such as skins of vegetables or fruits.

Peel To remove the outer covering of foods such as oranges or bananas.

Pinch The amount of a substance that can be held between the thumb and forefinger.

Pit To remove the seed from whole fruit.

Poach To cook in a hot liquid, being careful to retain shape of the food (poached eggs or fish, for example).

Punch down To deflate a risen yeast dough by pushing it down with the fist.

Puree To press food through a fine sieve or food mill; to blend in a blender or food processor to a smooth, thick mixture; also a thick sauce made from pureed vegetables.

Reconstitute To restore concentrated foods such as nonfat dry milk or frozen orange juice to their original form by adding water.

Reduce To decrease volume of liquid by rapidly boiling.

Rehydrate To restore water lost during drying by soaking or cooking dehydrated foods.

Roast To cook uncovered in an oven.

Sauté	To brown or cook in a small amount of fat.
Scald	To heat liquid to just below the boiling point; to pour boiling water over food or dip food briefly in boiling water.
Scallop	To bake food (usually cut into pieces) with a sauce or other liquid.
Score	To cut shallow slits on the surface of meat or food to increase tenderness or to prevent the fat covering from curling.
Sear	To brown the surface of foods quickly with high heat.
Simmer	To cook food over low heat in a liquid just below the boiling point (about 180 to 210°F or 80 to 99°C). Bubbles will form slowly and break apart just below the surface.
Steam	To cook food in steam over boiling water in a closed container.
Steep	To let a food stand in liquid below the boiling point of water to extract flavor or color.
Stew	To simmer food in a liquid.
Stir	To mix ingredients with a circular motion.
Stir-fry	To fry thinly sliced food quickly in only a little oil, stirring with a tossing motion.
Strain	To remove liquid from food using a strainer.
Toast	To brown by direct heat or in a hot oven.
Toss	To mix foods lightly with a lifting motion.
Warm	A temperature of 105 to 115°F (40 to 46°C).
Whip	To beat food such as cream, eggs, or gelatin rapidly to incorporate air and increase volume.

INTERNATIONAL CULINARY AND MENU TERMS

Áebleskiver	Danish pancake, often filled with apple bits, baked in a special pan so it forms a ball about two inches in diameter.
Á la carte	Menu in which each dish is ordered and priced separately.
Á la king	Style of presentation in which food is mixed with a seasoned cream sauce.
Á la lyonnaise	Cooked with onions, in the style of Lyon, France.
Á la mode	Made or served in a specific style, such as pie with ice cream.
Albóndiga	(Spanish) Meatball.
Al dente	(Italian) Cooking "to the tooth," a slightly firm but tender stage of cooking pasta.
Almandine	Containing or topped with almonds.

Andouille	(French) A reddish-pink sausage made from pork stomach and other ingredients.
Antipasto	(Italian) Assorted relishes.
Au gratin	Sprinkled with crumbs and/or cheese and baked until brown.
Au jus	Served with the natural juice or gravy.
Bagel	Doughnut-shaped roll that is cooked in water before being baked.
Baklava	Mid-Eastern confection made of multilayers of phyllo, honey, and chopped nuts.
Bannock	Bread of oat or barley flour, frequently unleavened, and baked in flat loaves; common in the northern British Isles. In New England it is a thin corn bread, baked on a griddle.
Bar-le-duc	Sugar syrup containing whole white or red currants and named for the town in Lorraine, France, where it originated. It is used as a topping for desserts or pastry filling.
Béarnaise sauce	(French) Sauce made with egg yolks and butter.
Béchamel	(French) A basic white sauce usually made with milk and sometimes with stock.
Beignet	(French) Variety of fritters fried in deep fat.
Bisque	(French) A thick, rich, creamy soup.
Blanquette	(French) A stew of light-colored meat or poultry in which the meat is cooked in water without browning.
Bleu or Au bleu	(French) Method of quickly cooking fish while still alive to guarantee maximum flavor and freshness; results in curled tail and bluish skin color.
Blini	(Russian, Polish) Small buckwheat crêpe usually served with caviar or sour cream.
Blintz	(Jewish) Thin crêpe with fruit or cheese filling, rolled and sautéed in butter.
Bombé	(French) Rich elaborate dessert made of ices, whipped cream, custard, and/or fruit frozen in a melon mold.
Borscht	(Russian, Polish) Thick soup made with a beet base, which may contain other vegetables, meat, and a topping of sour cream.
Bouillabaisse	(French) Thick, hearty soup made from an assortment of fish; originally made in the Mediterranean region of France.
Bouillon	(French) Broth or clear soup made from various meats, fish, poultry or vegetables.
Bouquet garni	(French) Combination of herbs, usually thyme, bay leaf, and parsley, tied together and used as flavoring.
Bratwurst	Mild German sausage, usually light gray in color.
Braunschweiger	(German) Smoked liver sausage.
Brioche	Rich classic French yeast dough, usually shaped with a topknot.

Bulgur	Cracked wheat originating in the Middle East; available as fine, medium, or coarse.
Buñuelos	Spanish or Mexican fried doughnut.
Café au lait	(French) Beverage composed of equal parts of strong coffee and heated milk poured simultaneously into the cup.
Café noir	(French) Black coffee.
Caldo	In Spanish literally a broth or gravy. In Mexico the term is broader and is used to describe soups or stews with meat and vegetables.
Canapés	Toasted, buttered bread spread or topped with appetizers, flavored spreads, vegetables or seafood.
Cannelloni	(Italian) Pasta dish made of squares of dough with meat or cheese filling, filled and baked in a sauce; crêpes sometimes are used instead of pasta.
Cannoli	(Italian) Cream-filled tubular pastry.
Capers	Pickled buds of a Mediterranean shrub with a tart, salty taste used as seasoning.
Cappuccino	(Italian) After-dinner coffee made from hot milk and espresso coffee, flavored with cinnamon.
Cassoulet	(French) Stew made from dried beans, assorted meats, and sausages.
Caviar	Black caviar is the eggs, or roe, of the sturgeon and is imported from Russia and Iran; red caviar is the roe of salmon.
Ceviche	Also Seviche. Raw fish marinated in lime juice.
Challah	Braided Jewish Sabbath bread.
Chantilly	(French) Containing whipped cream.
Chapati	Flat bread made from whole wheat, baked on a griddle; from India.
Chateaubriand	Thick piece of filet mignon usually prepared for two and served with Béarnaise sauce.
Chili con carne	Red beans, tomatoes, meat, and chili pepper seasoning; stew characteristic of southwestern United States.
Chili rellenos	(Mexican) Batter-coated green chili peppers stuffed with mild cheese.
Chitterlings	Also chittlins. Well-cleaned small intestines of hogs, cut in pieces and boiled, then fried and seasoned.
Chorizo	Hot Spanish or Mexican sausage
Chutney	Spicy-sweet relish originating in India, popular in England.
Cilantro	Pungent herb; also known as Chinese parsley or coriander leaves.
Compote	Stewed fruit, often based on dried fruit.

Consommé	Clarified double-strength brown stock.
Coq au vin	(French) Chicken cooked in red wine sauce.
Crêpes	(French) Very thin pancakes.
Croissants	(French) A rich, crescent-shaped roll or pastry.
Croutons	Small cubes of toasted bread used to garnish soup or salad.
Curry	Blend of many spices such as ginger, coriander, cardamom, cayenne, turmeric; originating in India.
Demitasse	Small cup of strong, black, after-dinner coffee.
Du jour	(French) "Of the day" used as soup du jour or the soup specialty available that day.
En Brochette	(French) Broiled and served on a skewer.
Enchilada	(Mexican) Tortilla rolled around meat or cheese filling and covered with sauce.
Entrée	In the United States, the main course; in France, the dish served before the main course.
Epice	(French) Spice.
Expresso	(Italian) Strong, black coffee made with steam.
Falaffel	(Israeli, Middle Eastern) Croquettes made from mashed chick peas fried in deep fat.
Feijoada	(Brazilian) Soup/stew of black beans, meat, and rice.
Fettuccine	(Italian) Narrow, flat noodles.
Fillet	Strip of boneless lean meat, fish, or poultry.
Fines herbes	Equal parts of freshly minced parsley, tarragon, chives, and chervil.
Fish sauce	This is used in Southeast Asian cuisine. The sauce is made from one type of fish and is often salty. Finger foods are dipped into the sauce.
Flambé	Served after flaming with brandy or other high-proof liqueur.
Flan	Baked custard with caramel sauce also called crème caramel; also open-faced pastry tart with fruit toppings.
Florentine	(Italian) Served in the manner of Florence with spinach or a spinach base.
Foie gras	(French) Specially fattened goose livers ground to a fine, smooth paste and seasoned.
Fondue Bourguignon	(Swiss) Small pieces of meat deep-fried at the table by each diner.
Fondue Suisse	(Swiss) Melted cheese, white wine and kirsch into which cubes of dry French bread are swirled.
Foo Yong	Chinese omelet with bean sprouts, meat, and fish.
Frijoles refritos	(Mexican) Cooked beans that have been mashed and fried.

Frittata	(Italian) Omelet with flavorful filling or topping.
Fritter	Batter-dipped food fried in deep fat or sautéed.
Gazpacho	(Spanish) Chilled soup composed of finely chopped raw tomatoes, cucumbers, onions and other vegetables in spiced tomato juice.
Gefilte fish	Also gefullete. German/Jewish stuffed or chopped fish.
Génoise	(French) Rich light sponge cake; often used as a base for more elaborate desserts.
Ghee	(India) A clarified butter.
Ginger root	Spicy tropical root used for flavoring in many cuisines; available fresh, dried and powdered, candied or pickled.
Ginseng	Thick root of plant highly prized in the Orient for its stimulating, refreshing qualities; plant originated in Korea.
Gnocchi	(Italian) Small, tender dumplings made of potatoes, flour, or semolina.
Goulash	(Hungarian) Beef stew liberally seasoned with paprika.
Guacamole	(Mexican) Smooth paste of avocado, tomato, onion, and green chilis used as an appetizer, salad dressing, and sauce.
Gumbo	Thick soup/stew characteristic of Creole cooking in Louisiana; also means okra.
Hassenpfeffer	(German) Spicy rabbit stew.
Hollandaise sauce	(French) Thick rich sauce made of egg yolks, butter, and lemon juice.
Hor d'oeuvres	Attractive small appetizers.
Huevos	(Spanish) Eggs.
Imbottite	(Italian) Stuffed.
Kabob	Also kebab (Middle Eastern) Skewered pieces of meat and vegetables roasted over flames.
Kasha	Russian buckwheat groats; also used in other East European cuisines as a staple.
Kim chee	(Korean) Spicy fermented cabbage.
Kippers	Split, dried, salted, smoked herring; popular English breakfast food.
Kuchen	(German) Cake or coffee cake; also specifically a flat round bread topped with cream cheese and sweetened fruit.
Kolachi	Also Kolachy. (Middle European) Yeast roll filled with cheese or preserved fruit.
Kugelhopf	(German) Yeast-raised dough usually baked in a Turk's head mold.
Lasagna	(Italian) Broad noodles layered with meat, cheese, and sauce.
Limpa	(Swedish) Rye bread flavored with orange rind.

Linguine	(Italian) A long, thin and narrow ribbon-like pasta.
Luau	Hawaiian feast.
Lucia buns	(Swedish) Sweet rolls served on St. Lucia Day, December 13.
Manicotti	(Italian) Large tubular noodles stuffed with cheese, meat, or vegetables and served with sauce.
Matzos	(Jewish) Thin flat pieces of unleavened bread eaten during Passover.
Menudo	(Mexican) Soup/stew made of tripe, hominy and green chilis.
Minestrone	(Italian) Thick vegetable soup.
Mocha	Type of Arabian coffee; also coffee or coffee/chocolate flavored.
Moussaka	(Middle Eastern) Layers of eggplant alternated with chopped lamb or beef.
Mousse	(French) Light delicate dish usually made with gelatin with sweet or savory fillings, depending on the use – sweet for dessert.
Nesselrode	Dessert or sauce made of candied fruit and chestnuts.
Nicoise	(French) In the style of Nice, meaning use of tomatoes, onions, garlic, capers and olive oil.
Paella	(Spanish) A rice casserole with chicken, seafood, vegetables, and seasonings.
Pannetone	(Italian) Wine-flavored yeast bread filled with candied fruit, raisins, and nuts.
Pareve	(Jewish) In kosher cuisine, foods such as fruits, vegetables, and eggs that may be eaten with either meat or dairy foods.
Pasta	All forms of macaroni products.
Pâté	(French) Ground, seasoned meat or meats served with or without pastry covering.
Petits fours	(French) Small fancy iced cakes.
Phyllo	Also Fila, brik, malsouka, yukka. A tissue-thin pastry which is used in the Middle East.
Pilaf	Also pilau (Middle Eastern). Browned rice, barley, or cracked wheat variously seasoned.
Pirozhki	(Russian) Small meat-filled pastries used as appetizers.
Pita	Flat bread with a pocket inside that may be stuffed with meat and/or salad.
Poi	(Hawaii and South Pacific) Starchy, bland paste made from fermented taro root.
Polenta	(Italian) Corn meal porridge.
Quesedilla	(Mexican) Cheese wrapped in a tortilla and heated.
Quiche	(French) An unsweetened custard frequently containing cheese and usually baked in a pastry shell.

Ragôut	(French) Stew made with meat, poultry, or fish, and sometimes vegetables.
Ravioli	(Italian) Filled pasta dough that is boiled and served with a sauce.
Rigatoni	Large fluted tubes of Italian pasta, often stuffed and covered with sauce.
Rijsttafel	Literally "rice table"; Dutch-Indonesian dinner with rice and an assortment of spicy foods.
Risotto	(Italian) Wide variety of rice dishes.
Roquefort	(French) Blue-veined, pungent cheese; authentic Roquefort comes only from a certain area of France and bears the trademark of a small red outline of a sheep on the label.
Roux	(French) Mixture of fat and flour used to thicken soups, sauces, and gravies.
Salsa	Spanish for sauce. In Mexico a mixture of chopped fresh tomatoes, peppers, and onions.
Sashimi	(Japanese) Raw fish served with dipping sauces, usually an appetizer.
Sauerbraten	(German) Pot roast of beef that has been marinated several days in spicy vinegar sauce.
Smorgasbord	(Scandinavian) Literally "sandwich table"; large assortment of hors d'oeuvres, cold salads, meats and desserts.
Smorrebrod	(Danish) Open-face sandwiches, tastefully arranged and garnished.
Soufflé	(French) A delicate, baked dish with a base of very thick white sauce and egg yolks blended with whipped egg whites.
Spätzle	Also Spatzen. Small German egg dumplings.
Spumoni	(Italian) Rich ice cream containing candied fruit and nuts.
Stollen	(German) Pastry or yeast coffee cake containing dried and candied fruit and nuts; associated with Christmas.
Strudel	(Austrian) Thin leaves of flaky pastry filled with fruit, cheese, meats or vegetables based on a stretched dough; characteristic of central Europe.
Stroganoff	(Russian) A meat mixture with noodles, onions, mushrooms, and a sauce with a sour cream base.
Subgum	In Chinese cuisine, meaning mixture of vegetables.
Sukiyaki	(Japanese) Sautéed beef and vegetables; usually cooked at the table.
Sushi	(Japanese) Appetizer composed of vinegared rice topped with raw fish or wrapped in laver (seaweed).
Tàble d'Hôte	(French) In restaurants a fixed price for all the courses of a meal; opposite of à la carte.
Taco	(Mexican) Fried tortilla folded around various fillings.
Tahina	(Middle Eastern) Oily paste prepared by grinding sesame seeds.

Tamale	(Mexican) Ground corn paste shell wrapped around meat or poultry filling and steamed.
Tempura	(Japanese) Batter dipped, deep-fried food of any type.
Teriyaki	(Japanese) Soy sauce marinade used to flavor meat and fish.
Timbale	A custard which is cooked in a drum-shaped mold; a pastry shell filled with a creamy mixture.
Tofu	Soy bean curd used in oriental cooking.
Torte	A rich cake usually made of eggs, chopped nuts, and crumbs.
Tortilla	(Mexican) Thin, unleavened corn bread; in Spain an omelet.
Tostada	(Mexican) Layered salad composed of tortilla, refried beans, shredded lettuce, ground meat, and grated cheese.
Vermicelli	(Italian) Very narrow type of spaghetti.
Vichyssoise	(French) Cream soup made with leeks and potatoes; usually served cold.
Vinaigrette	(French) Marinade or salad dressing of oil, vinegar, pepper, and herbs.
Vol-au-vent	(French) Flaky or puff pastry made into a shell.
Wiener schnitzel	(Austrian) Vienna style breaded, sautéed veal cutlet.
Won ton	(Chinese) Squares of noodle dough that are wrapped around meats and vegetables and cooked in boiling water.
Yorkshire pudding	(English) A batter pudding baked in the drippings of roasting meat.

Milk and Milk Products

Milk is the main food for young mammals and is used in countless ways as food in fresh, fluid form and food products. The term milk usually implies cow's milk in the United States. All milk and milk used in products should be pasteurized to destroy pathogenic bacteria (disease-causing bacteria) or ultra-pasteurized and meet local ordinances and state standards as well as federal specifications.

BUTTER

Butter is made from sweet or sour cream and contains not less than 80 percent milkfat. A lactic acid culture may be added to the cream for a short "ripening" period to develop desirable aroma and flavors before the cream is churned. The addition of annatto, a colorant derived from the annatto plant and salt is optional.

Butter that has been officially graded by the U.S. Department of Agriculture for sale on the retail market bears a shield on the package with a letter grade that indicates the quality of the butter at the time of grading. Grades AA and A are available at the retail level. Grade B is usually used as an ingredient in manufactured applications, rather than sold as a product. Occasionally the equivalent numerical flavor score of butter is shown under the grade shield. Grades for butter depend on quality factors of flavor, body, texture, color, and salt. The U.S. grade and score relationships are: AA or 93 score, A or 92 score, B or 90 score, and C or 89 score.

Whipped butter has been stirred or whipped to incorporate air or some inert gas to thereby increase the volume and make the butter easier to spread.

STORAGE AND USE

- To store butter, leave it in its original package and keep in the food compartment of the refrigerator or in the freezer.
- Place partially used portions of butter in a covered dish, refrigerate, and use up within a few days.

- Wrapped in moisture-vapor-proof packaging material, butter can be frozen for up to nine months at 0°F (-18°C).
- To measure unwhipped butter, press it firmly into individual measuring cups or spoons and level with the straight edge of a spatula. Or measure butter according to these equivalents: 1 pound = 2 cups = 4 sticks = 32 tablespoons (454 g = 473 ml).
- Whipped butter should not be substituted in most recipes for regular butter or margarine because the volumes are not equivalent.
- To substitute other fats in recipes that call for butter, allow for each cup of butter 1 cup (237 ml) margarine or $7/8$ cup (207 ml) hydrogenated shortening or lard plus $1/2$ teaspoon (2.5 ml) salt.

CHEESE

Cheese comes in many forms and in a wide variety of flavors and textures.

NATURAL CHEESE

Natural cheese is made from cow's, sheep's, or goat's milk or cream and is cured or aged for a specific period to develop flavor. It is prepared by precipitating proteins in milk and separating the curd or solid portion from the whey or watery portion. Natural cheese may be classified by texture or consistency and the degree or kind of ripening. Harder cheeses keep longer. Classifications include:

Very hard cheese—Parmesan, Romano, Sap Sago, bacteria-ripened.

Hard cheeses—Cheddar, Colby, Colby Jack, Swiss, and Provolone, bacteria-ripened.

Semisoft to hard cheeses—Edam, Baby Swiss, Gouda, and Feta, bacteria-ripened; Gjetost, and Mysost, whey cheeses, unripened.

Semisoft cheeses—Blue (Bleu), Gorgonzola, Roquefort, and Stilton, mold-ripened; Brick, Monterey Jack, and Muenster, Port du Salut, bacteria-ripened; mozzarella, unripened.

Soft cheeses—Brie and Camembert, mold-ripened; Limburger, Bel Paese, bacteria-ripened; cottage cheese, cream cheese, Neufchatel, and ricotta, unripened.

Grades have been established for cheese such as Cheddar, Colby, Monterey and Swiss based on specifications for flavor, odor, body, texture, finish, appearance, and color with additional specifications applicable to cheese of different ages. The quality rating of cheese is seldom given in the retail market. Swiss cheese is rated for eye formation, as well as for flavor, body, and texture.

PROCESS CHEESE AND PRODUCTS

Process cheese is a blend of one or more natural cheeses that have been processed using heat to inactivate bacteria and enzymes. Added emulsifiers enhance blending qualities. Included in this group are pasteurized process cheese, process cheese food, process cheese spread and coldpack cheese, which is not heated.

Pasteurized process cheese has up to 3 percent emulsifying salts such as sodium citrate or disodium phosphate and may contain milk, cream, or whey added to the natural cheese. The maximum moisture content is 40 percent, which is higher than the same type of natural cheese.

Cheese food is a product made from a mixture of one or more varieties of cheese with added solids, salt, and up to 3 percent of an emulsifier, all of which are mixed with the aid of heat. The moisture content of a pasteurized cheese food is about 4 percent higher than that permitted for process cheese; the milkfat content is lower. Fruit, vegetables and meats may be added.

Cheese spreads are similar to cheese foods except that a stabilizer is used, moisture content is about 8 percent more than in process cheese, and milkfat content is lower. Spreads may be flavored with pimiento, olives, pickles, onion, or other added ingredients. Fruit, other vegetables and meats may be added.

Cold pack cheese or club cheese is a blend of natural cheeses prepared without heat. Stabilizers, spices or smoke flavoring are often added. Softer in texture than natural cheese, cold pack cheese spreads readily.

REDUCED-FAT AND LOW SODIUM CHEESE

Reduced-fat cheese may be a natural cheese or a processed cheese. Natural reduced-fat cheese is made from reduced-fat milk and contains less fat and fewer kilocalories than its counterpart. A reduced-fat cheese generally contains more protein and calcium than its full-fat counterpart.

"Light" pasteurized process cheese is made by blending reduced-fat natural cheese with an emulsifier to increase keeping quality and improve melting consistency.

Fat-free cheeses are made from skim milk and cheese cultures. Fat-free ricotta, cottage cheese, and other cheese products are available. More fat-free and reduced cheese products are currently under development.

Low sodium cheese contains 140 mg or less sodium per serving. Some varieties are available with two-thirds to one-half less sodium than their regular counterparts.

Imitation cheeses are made from cows' milk that does not meet Federal Standards of Identity for composition. They may be made from filled milk or mixtures of sodium and calcium caseinate, emulsifying salts, flavorings, and vitamins and minerals.

STORAGE AND USE

- All cheese keeps best at refrigerator temperatures of 35 to 40°F (2–4°C). Store in the original wrap or container until opened, then rewrap cheese tightly in plastic wrap or place in a clean, airtight container.

- Cheese should be used by the date listed on the container or wrapper.

- Uncreamed dry curd cottage cheese may be frozen in appropriate freezer containers. It will keep well for about a month. Textural changes may occur.

- Most hard cheeses can be frozen for six to eight weeks, but freezing results in a crumbly texture, which is difficult to slice. To freeze cheese, tightly wrap pieces of one pound (454 g) or less and about an inch (2.5 cm) thick in freezer wrap or in moisture-proof, air-tight packaging and freeze at 0°F (-18°C). Thaw cheese in the refrigerator and use within a few days after thawing.

- For full flavor and aroma serve cheese at room temperature or 20 to 60 minutes after removal from the refrigerator. The exceptions are soft, unripened cheeses such as cottage, cream or Neufchatel cheeses which should be served chilled unless being used as an ingredient in omelets, casseroles, and other hot dishes.

- Cooking temperatures for cheese should be low to prevent stringiness and toughness. High temperatures cause cheese to become leathery, lose flavor and separate fat.

- Cheese that is thoroughly chilled grates or shreds more easily than cheese at room temperature.

- When mold is found on cheese, trim at least a 1-inch section of cheese away from the affected area and discard. Use the remaining cheese immediately.

- Reduced-fat cheeses can be added to casseroles, salads or sandwiches. The low-fat cheese does not "string" and melt as its full-fat counterpart does.

CREAM

Cream is the fat portion of milk that rises to the surface when nonhomogenized milk is allowed to stand. Cream may be separated from milk by centrifugal force. To meet Federal Standards of Identity, cream must have not less than 18 percent milkfat. Emulsifiers, stabilizers, sweeteners, flavorings or other optional ingredients may also be added.

Half-and-half is a mixture of milk and cream with 10.5 to 18 percent milkfat. Half-and-half is usually homogenized.

Light cream is coffee or table cream that is sometimes homogenized. It usually has 20 percent milkfat, but the fat content can range between 18 to 30 percent milkfat.

Light whipping cream contains between 30 and 36 percent milkfat and whips up satisfactorily but does not freeze successfully.

Heavy cream or whipping cream contains 36 to 40 percent milkfat and is not homogenized. As a foam it tends to be more stable than light whipping cream foams.

Pressurized whipped cream is whipping cream packed in aerosol cans under pressure. It is made from a mixture of cream, sugar, stabilizers, emulsifiers, flavors and a gas-forming substance such as nitrous oxide. State regulations specify kind of cream and minimum fat content.

Dairy sour cream is sweet cream to which a lactic acid culture has been added. Sour cream must comply with individual state requirements for minimum milkfat of the cream used (usually 18 to 20 percent).

Sour half-and-half is made with half-and-half rather than cream, and the acidity is not less that 0.5 percent. Products that resemble sour cream are sour half-and-half dressing and sour cream dressing in which butter or other dairy ingredients replace the cream or milk.

Reduced-fat sour cream is made like dairy sour cream, but skim milk has been added to reduce the fat content. There is also fat-free sour cream available with thickening agents.

Sour cream dressing or sour half-and-half dressing resembles sour cream and sour half-and-half except some of the cream and/or milk is replaced by other dairy ingredients such as butter.

STORAGE AND USE

- Cream should be kept in its original container and refrigerated at 40°F (4°C) or lower until used.

- Cream has a pull date on the package label. The pull date indicates when the product should be withdrawn from retail sale. This date allows for additional storage life when the consumer's refrigerator is maintained at 35–40°F (1–4°C). Ultra-pasteurized cream will keep several weeks in the refrigerator. Once opened, it should be treated like pasteurized cream and used within a few days.

- In recipes using sour cream, add the sour cream at the end of the cooking period or combine with flour or condensed soup to help prevent separation or curdling.

MILK

Pasteurized milk has been subjected to temperatures that ensure destruction of pathogenic (disease-causing) microorganisms, some spoilage bacteria and natural milk enzymes. This results in a prolonged refrigerated shelf life. However, it has not been sterilized. Milks pasteurized by vat, batch (LTLT (long-time-low-temperature), 145°F, 63°C 1800 seconds), high-temperature-short-time (HTST), higher-heat-shorter time (HHST), or ultra-pasteurization must be refrigerated. Milk pasteurized by ultra-high temperature (UHT) (280°–320°F, 138–150°C for 2–6 seconds) and packaged in aseptic containers is shelf stable for about three months. When opened, it must be refrigerated.

FLUID MILKS

Fluid milks are classified primarily by the milkfat content, nonfat milk solids and added nutrients. Vitamin D is regularly used to fortify most fluid milks—400 International Units (IU) per quart (946 ml). Fluid lowfat milk products also may be fortified with vitamin A, which is removed with the milkfat. The vitamin A is added at 2000 IU per quart (946 ml). Most fluid milks with fat are **homogenized** to stabilize the dispersion of milkfat globules throughout the milk, so the fat does not rise to form a cream layer. The milk is forced under pressure through small apertures, which break the fat globules into smaller ones.

	Milkfat %	Nonfat Milk Solids %	Nutrient Added
Whole milk	≥3.24	≥8.25	D
Lowfat milks			
Skim/nonfat	≤0.5	+8.25	A, D
	1.0	+8.25	A, D
	1.5	+8.25	A, D
	2.0	+8.25	A, D

Flavored milk includes a variety of chocolate or cocoa flavored products. Other flavors such as strawberry, maple and coffee may be used for flavored milks or drinks. The caloric content of flavored milks or drinks varies according to the amount of milkfat and composition of other added ingredients.

Kind of	Milk	Other Flavor	Additives
Chocolate milk	Whole	Chocolate	Sweetener
Chocolate-flavored milk	Whole	Cocoa syrup	Sweetener
Chocolate drink	Skim/lowfat	Chocolate	Sweetener
Chocolate-flavored drink	Skim/lowfat	Cocoa syrup	Sweetener

CULTURED AND SOURED MILKS

Cultured buttermilk is obtained by treating pasteurized skim or part-skim milk with a suitable culture of lactic acid bacteria. The lactic acid produced gives the characteristic acidity and increased thickness of the product. Buttermilk originally was a byproduct of butter making.

Sour milk is milk soured naturally or artificially by the action of lactic acid bacteria or artificially by the addition of vinegar or lemon juice. Milk that has been soured to the stage that a firm curd has formed is called clabber. The whey and curds have not separated.

Acidophilus cultured milk is prepared by the addition of a *Lactobacillus acidophilus* concentrated culture to lowfat or skim milk that has been pasteurized, homogenized, and chilled. The rapid chilling halts bacteria growth retaining the sweet flavor and consistency of the milk.

Bifidobacteria milk, as produced and marketed in the United States, is a nonfermented, lowfat milk containing 1 percent acidophilus/bifidobacteria culture. Bifid-amended dairy products are in great demand in Japan and Europe. Products fermented with Bifidobacteria culture have a mild, acidic flavor, similar to that of yogurt. The culture is added during fermentation or to the finished fermented or fresh product prior to shipment.

NOTE: The fat percentages used with milk products are expressed as a percent of weight not caloric content.

YOGURT

Yogurt is a creamy, custard-like cultured milk product made by fermenting whole, low-fat or skim milk with a bacterial culture such as *Streptococcus thermophilus* and *Lactobacillus bulgaricus*. The minimum amount of milk solids is 8.25 percent with an average of 15 percent solids and 0.5 percent acid. The milkfat standards for lowfat yogurts range from 0.5 to 3 percent; nonfat yogurt contains 0.5 percent or less.

Plain, flavored-containing-no-fruit and flavored-with-fruit yogurts are available. Coffee, vanilla, and lemon are examples of flavored yogurt containing no fruit.

If yogurt contains a fruit it may be **sundae-style** with fruit on the bottom of the container topped with plain or flavored yogurt or **Swiss- or French-style** with fruit blended throughout plain or flavored yogurt.

Frozen yogurt is prepared by adding suitable stabilizers, nonfat milk solids, sweeteners, fruits, and juices to cultured yogurt before freezing. It may be soft or hard. Federal standards have not been established for frozen yogurt.

Low-lactose yogurt has about 60 percent less lactose than milk, making this product suitable for lactose-intolerant people.

CONCENTRATED AND DRIED MILK PRODUCTS

Evaporated and evaporated skimmed milk has about 60 percent of the water removed. They may be fortified with vitamins D and/or A. Whole milk concentrates are homogenized. The milk is then sealed in cans and sterilized. According to federal standards, evaporated milk must contain at least 7.5 percent milkfat and not less than 25.5 percent total milk solids. Evaporated skimmed milk must contain not less that 0.5 percent milk fat and a minimum of 20 percent total solids and be fortified with 125 IU of vitamin A per fluid ounce (30 ml). The composition of fats and solids of evaporated milks mixed with an equal volume of water is slightly above the average of fresh milks.

Sweetened condensed milk and skimmed milk are the products resulting from the evaporation of about half the water from whole or skimmed milk, and the addition of refined cane and/or corn sugar in amounts (44 percent) sufficient for preservation. Therefore, this form of canned milk has not been sterilized. The high sugar adds to the browning of the milk when heated. The high concentration of milk protein allows thickening with the addition of acid or heat to make toppings or desserts.

Dried whole milk is similar to fresh whole milk but it is packed in vacuum-packed or gas-packed containers to prevent deterioration. It contains not less than 26 percent milk fat and not more than 4 percent moisture. U.S. grades are Extra and Standard with dry whole milk also having a Premium grade.

Nonfat dry milk is the product resulting from the removal of fat and water from pasteurized milk. It contains not over 5 percent moisture and 1.5 percent milkfat by weight. Nonfat dry milk contains the same relative proportions of proteins, minerals and water soluble vitamins of fluid nonfat milk and is fortified with vitamins A and D. Most nonfat dry milk is instantized to form rather coarse, creamy-white, free-flowing particles that dissolve readily in water.

SPECIALTY MILKS

Multivitamin and multimineral milks to which multivitamin and multimineral concentrates are added are sold in some localities.

Lactase-treated milk is processed milk to which lactase, and enzyme, is added. The lactase hydrolyzes lactose to glucose and galactose. Lactose-intolerant individuals can use lactase-treated milk or purchase lactase tablets to add to milk.

Low Sodium Milk is a product which has about 95 percent of the sodium replaced with potassium by an ion-exchange method resulting in about 6 mg of sodium per cup (237 ml) as compared to 120 mg in regular milk. Both fresh or canned low sodium milk may be available.

Certified Milk is a costly and sanitary product produced according to American Association of Medical Milk Commissions' *Methods and Standards for the Production-of Certified Milk*. Its sale is not permitted in all states. The milk may be raw, pasteurized, homogenized and/or fortified with vitamin D. Raw certified milk may cause foodborne illnesses.

Kefir and koumiss are fermented milk beverages commonly produced in Southwestern Asia and similar in taste to yogurt. The are prepared by using kefir grains which contain bacteria and a lactose-fermenting yeast held together by layers of coagulated protein. Acid production is controlled by the bacteria, while yeast produces alcohol and some carbon dioxide. The final concentration of lactic acid is 0.8 percent, and alcohol may be as high as 1 percent.

Eggnog is made from pasteurized milk or cream, pasteurized eggs, sugar and flavored with vanilla. It is available during the holiday season.

STORAGE AND USE

- Use milk in order of pull date found on the container. Yogurt keeps about 10 days after the pull date stamped on the carton.

- Unopened packages of any shelf-stable milk should be stored in a cool, dry place. After opening, the unused milk must be refrigerated.

- Evaporated milks may be substituted for other milks. Sweetened condensed milk, however, is not a satisfactory substitute because of the high sugar content that affects flavor and texture.

- To measure dry milk, pour from package or spoon lightly into individual measuring cup, heaping to the brim. Level with straight edge of spatula or knife. Shaking the cup to level the powder tends to pack it down and gives an inaccurate measure.

- In recipes, dry milk may be added either to dry or liquid ingredients unless the recipe specifically calls for reconstitution.

- For best flavor, milk for drinking should be served icy cold from the refrigerator. Store milk in its original container.

- Use of acidophilus milk in cooking will destroy the culture.

FROZEN DAIRY PRODUCTS

Ice cream is made from a pasteurized mixture of milk, cream, sugar, stabilizers, flavorings, and sometimes eggs. Coloring may also be added. As the mixture is frozen, it is whipped to approximately 80 to 100 percent more than its original volume. The finished product usually weighs about 4.5 pounds per gallon (2 kg/3.71) and contains not less than 1.6 pounds (726 g) of food solids per gallon (3.71). The milkfat content ranges from 8 to 14 percent, usually 10 to 12 percent for plain ice cream, although some special ice creams may be as high as 20 percent. Federal standards require that plain ice cream contain at least 10 percent milkfat and 20 percent total milk solids by weight. Ice cream with chocolate, nuts, or fruits must contain at least 8 percent milkfat and 16 percent total milk solids.

Frozen custard and French custard ice cream are products with high levels of egg yolk solids cooked as a custard before freezing.

Ice milk is a product made in the same manner and with the same ingredients as ice cream but the ingredients are present in different proportions. The milkfat content ranges from 2 to about 7 percent, and the total milk solids must be at least 11 percent according to the federal standard. It may be either soft- or hard-frozen drawn from the freezer.

Imitation ice cream and ice milk are currently available. Products which do not meet the definition of ice cream must be labeled imitation. Mellorine has the butterfat replaced by a suitable animal or vegetable fat. Parevine contains no dairy ingredients. Diabetic products may contain sorbitol instead of sugar to lower the freezing point and to bind water. Low-calorie ice cream may use non-nutritive sweeteners, or high-intensity nutritive sweetners with bulking agents. Low-calorie or noncaloric fat substitutes can be used in products. One is a microparticulation of egg white and/or milk. This product, Simplesse® is used in some ice cream, yogurt, sour cream and cream cheese.

Sherbet is a frozen product made of a pasteurized mixture of sugar, milk solids, stabilizer, flavorings such as fruit, fruit juices, or extract, and water. Milk fat content of sherbet is 1 to 2 percent, and the total milk solids content is 2 to 5 percent.

Refrigerated puddings are made with whole or skim milk, packaged and held at refrigeration temperatures in the grocery store and at home.

STORAGE AND USE

- Store ice cream in a tightly closed container at 0°F (-18°C) for up to two months.
- For easy serving, frozen dairy products should be removed from freezer to refrigerator 10 to 20 minutes before serving.
- Ice crystals can be minimized by covering the product with plastic wrap before closing. Small crystals also are retained if the mixture is not allowed to freeze and thaw (or the temperature stays stable).

WHIPPING PROPERTIES

Cream—When whipped, cream increases two to three times in volume depending on the type of beater used. The fat content should be 30 to 40 percent. Pasteurization and homogenization decrease foam volume and stability. Cream chilled for 48 hours has increased foam volume and stability.

- To whip cream, chill utensils (50°F, 10°C) before use. Keep utensils and cream cold during preparation. Choose a deep bowl just wide enough for beater. Add sugar after whipping, never before. Beat rapidly until stiff, but do not over beat or butter flakes will be formed.

- Whipped cream should be kept chilled at 50°F (10°C) or below until used. The foam is stable for about one hour.
- Whipped cream freezes well, particularly in individual serving portions. Once the portions are frozen, wrap in moisture-proof, airtight packaging and keep frozen until used.

Evaporated milk—Evaporated milk, when whipped, increases two to three times in volume. The foam is less stable than whipped cream unless the milk is made acid or supplemented with gelatin.

- Undiluted evaporated milk will whip best if it is first chilled below 32°F (0°C) until fine ice crystals form.
- To stabilize the foam, increase the acidity by adding 1 tablespoon (15 ml) lemon juice or vinegar for each cup (250 ml) of undiluted evaporated milk. Whip the milk and then fold in the juice or soften 1/2 teaspoon (2.5 ml) unflavored gelatin in 2 teaspoons (10 ml) cold water and dissolve in 1 cup (250 ml) scalded undiluted evaporated milk. Chill as above, whip, sprinkle with 1/4 cup (60 ml) powdered sugar and fold in.

Nonfat dry milk—Nonfat dry milk triples in volume when whipped and is stable for several hours.

- Use 1/3 cup (80 ml) cold water to 1/2 cup (125 ml) instant nonfat dry milk. Chill and whip until the mixture is thick enough to stand in soft peaks. Add 1 tablespoon (15 ml) lemon juice and continue to whip until stiff peaks form. Then beat in 2 to 4 tablespoons (30 to 60 ml) sugar.
- Whipped nonfat dry milk should be refrigerated and kept chilled until served.

Meat

Meat is generally defined as the edible portions of cattle, swine, and sheep. The major physical component of meat as it appears in the market is muscle tissue (lean). Depending on the cut of meat, adipose tissue (fat), connective tissue, and skeletal tissue (bone) may also be present.

The most abundant component of meat is water, which ranges from 45 to 75 percent. The amount of water in the tissues, which decreases as the animal ages and deposits fat, affects the texture, color, and flavor of the muscle.

High quality proteins are abundant in all muscle tissue. Different proteins behave in various ways during food preparation. Some toughen; others become more tender during cooking. The amount of fat varies with the species, the particular animal, and the cut.

The amount of adipose tissue differs widely among carcasses. As the animal grows and ages, subcutaneous fat is deposited just beneath the skin and abdominal fat is deposited around the kidney and in the pelvic area. Then fat is deposited between the muscles (intermuscular or seam fat). Finally, fat is deposited within the muscles (intramuscular fat or marbling).

Other important nutrients in meat that occur in significant amounts are the minerals iron, phosphorous, zinc, and most of the B vitamin complex (thiamin, riboflavin, niacin, B_6, and B_{12}). Vitamin B_{12}, essential for growth and the synthesis of DNA, occurs naturally only in animal foods.

Meat is marketed in many forms: fresh, frozen, freeze-dried, canned, cured, smoked, dried, and cured and smoked. Many meat items are available both raw and cooked.

FRESH MEAT

BEEF

The major sources of beef on the retail market are from:

Steers—male animals that have been castrated as calves: Steers (15 to 24 months old) account for more than 2/3 of the high grade beef on the market.

Heifers—young female animals that have not borne calves. Heifers produce high grade meat.

Cows—female bovines that have borne one or more calves.

Bullocks—young bulls under two years of age that have begun to develop secondary characteristics of a bull.

VEAL is a form of immature bovine animals under four months of age. It contains little fat.

CALF is meat from immature cattle, about four to nine months old. Calves weighing 400 to 800 pounds (181 to 363 kg) at the time of slaughter are classified as "baby beef."

LAMB

Meat from sheep under one year of age accounts for about 90 percent of the lamb on the market. Lamb is designated as:

Spring Lambs—lambs from three to seven months, weighing 70 to 120 pounds (32 to 54 kg).

Yearlings—ovines approximately one to two years of age.

Mutton—meat from sheep that are older than two years.

PORK

Most pork is from young swine of either sex less than one year of age. Most of the pork on the retail market is from:

Pigs—porcine animals that are less than four months of age and/or weigh less than 120 pounds (54 kg).

Hogs—porcine animals that are more than four months of age and weigh more than 120 pounds (54 kg).

Barrows—male porcine animals castrated when young.

Gilts—female porcine animals that have not had offspring or have not reached an evident stage of pregnancy.

Although most beef, lamb, and veal cuts are sold fresh, much pork is also available in cured or cured and smoked forms. See Table 4.1 for the location and bone shape of various retail cuts of beef, veal, lamb, and pork.

VARIETY MEATS

Organ meats from beef, veal, pork, and lamb are excellent sources of many essential nutrients. They include: brains, heart, kidneys, liver, tongue, tripe (stomach tissue), and sweetbreads (thymus).

PROCESSED MEAT

Processed meat may be defined as meat that has been changed by any mechanical, chemical, or enzymatic treatment, which alters the taste, appearance, or keeping quality of the product. It is made from wholesome meat that has been inspected at every stage of production. Major categories are:

SAUSAGE

Seasoned chopped meat may be sold in bulk or as stuffed links, with casings of natural or synthetic materials. Trimmings and some of the less popular meat cuts from pork, beef, veal, lamb, poultry, or a combination of these meats may be used in the processing of sausage.

More than 200 varieties of sausages and cold cuts are marketed in the United States. Some sausages are made from meat that is cured, smoked, cooked, or treated by a combination of these processes. Others may be finely or coarsely textured, fermented, dried, or semi-dried. Sausages are classified as:

Fresh—sausages made from ground meat and not cured, smoked, or cooked. They must be cooked before serving. Examples are fresh pork sausage, fresh bratwurst, chorizo, and Italian-style sausage.

Uncooked and Smoked—sausages that are smoked after forming and must be cooked before serving. Examples are smoked pork sausage, kielbasa, and smoked country-style sausage.

Cooked—sausages that have been cured and cooked. They are ready-to-eat, but may be heated before serving. Blood sausage, precooked bratwurst, liverwurst, and braunschweiger are some examples.

Cooked and Smoked—products made from meat that has been cured, formed into sausages, cooked, and lightly smoked. They are ready-to-eat, but may be heated before serving. Examples are cotto salami, bologna, frankfurters, and Vienna sausage.

Dry and Semi-Dry—sausages that may be smoked or unsmoked. Processing always includes curing and generally involves cooking at the plant. Carefully controlled fermentation acts as a preservative and provides the distinctive flavors. Hard salami is usually a dry sausage, while summer sausage is semi-dry. Cervelat, thuringer, pepperoni, and mortadella are examples of these ready-to-eat sausages.

TABLE 4.1

Identification of Meat Cuts[1]

Location	Beef	Veal	Lamb	Pork
Breast and forelegs	Brisket Shank Cross Cuts Short Plate	Breast Riblets Shank Cross Cuts	Breast Breast, rolled Riblets Spareribs Shanks	Spareribs Bacon Salt Pork
Shoulder/Chuck	Arm Steak or Pot Roast Blade Steak or Roast Chuck Eye Steak or Roast 7-bone Steak or Pot Roast Short Ribs Shoulder Steak or Pot Roast	Arm Steak or Roast Blade Steak or Roast	Arm Chops or Roast Blade Chops or Roast Shoulder Roast Neck Slices	Arm Steak or Arm Picnic Roast Blade Steak Blade Boston Boston Butt
Rib	Rib Steak or Roast, small end Rib Roast, large end Rib Eye Steak or Roast	Rib Chops or Roast Crown Roast	Rib Chops or Roast Crown Roast	Rib Chops Center Rib Roast
Loin (Short Loin)	Cube Steak Top Loin Steak or Roast Tenderloin Steak or Roast (Filet Mignon) T-Bone Steak Porterhouse Steak	Loin Chops or Roast Loin Kidney Chops Top Loin Roast	Double Loin Chops Loin Chops or Roast	Loin Blade Chops or Roast Loin Chops Center Loin Roast Tenderloin Top Loin Chops or Roast Canadian Bacon Country-Style Ribs Backribs
Sirloin	Sirloin Steak flat, pin, round or wedge bone Sirloin Roast Top Sirloin Steak	Sirloin Steak or Roast	Sirloin Chops or Roast	Sirloin Chops or Roast Sirloin Cutlets
Leg (Ham/Round)	Bottom Round Steak or Roast Eye Round Steak or Roast Top Round Steak or Roast Rump Steak Rump Roast Round Tip Steak or Roast	Round Steak Roast Rump Roast Cutlets	Leg Leg Slice or Roast	Ham Steak Leg (Fresh Ham) Roast Leg (Fresh Ham) Center Slice
Flank	Flank Steak			

[1] Location and bone of retail cuts of meat are often clues to identification.

Specialty Meats—meats that are cured and fully cooked during processing and are ready-to-eat. Loaf examples are honey, minced ham, old-fashioned, olive, pepper, and pickle and pimiento.

CURED WHOLE MUSCLE CUTS

Curing and drying are among the oldest methods of preserving meats. Salt, sugar, nitrites, and nitrates may all be used in the curing process. Salt acts as a preservative and a flavor enhancer; sugar also improves the flavor. Today, nitrites are the primary curing ingredient used in meat products. Nitrites develop the characteristic red-pink color in the lean parts of meat, impart a typically cured flavor, and inhibit the growth of a number of food poisoning and spoilage microorganisms, including *Clostridium botulinum*. Cured whole muscle cuts include:

Ham—the hind leg of pork that has been cured and smoked or cured and canned. One of the most highly valued pork products, ham can be produced three ways: bone-in, semi-boneless, and boneless. There are also cooked and uncooked varieties. To be labeled "fully cooked," hams must be cooked to an internal temperature of 148°F (65°C) at the processing plant. No additional heating is necessary for fully cooked hams. To serve warm, however, ham should be cooked to 130°F to 140°F (55 to 60° C) for optimum palatability. Hams that are labeled "cook-before-eating" must be cooked to an internal temperature of 160° F to 170° F (71 to 77° C) before serving.

Country-style hams—specialty hams that have been dry-cured, smoked, and aged to provide a distinct flavor and texture. Before baking, these hams require soaking or simmering.

Corned (Cured) Beef—prepared from brisket or round. Formerly, "corning" referred to preserving meat by sprinkling it with grains ("corns") of salt. Today, corned beef is cured with a pickle solution of water, salt, sugar, nitrite, and spices.

Bacon—produced primarily from pork bellies. Beef bacon is made from boneless beef short plate, which is similar to the belly in pork. Canadian-style bacon is made from the top loin muscle of pork, usually from heavier hogs. Both bacon and Canadian bacon are cured and smoked.

Dried Beef—made from the round that has been cured, lightly smoked, and dried.

Pastrami—made from beef brisket, plate, or top round. After dry-curing with salt, the beef is rubbed with a paste of garlic powder, ground cumin, red pepper, cinnamon, cloves, and allspice. It is then smoked and cooked.

Pork Shoulder—can be divided into two major cuts: Picnic (arm) and Boston (blade). They can be prepared by the same methods used for hams. Shoulder cuts are usually priced more economically than leg cuts, due to differences in palatability and yield of edible portion.

RESTRUCTURED MEAT

Restructured meat products are generally made from flaked, ground, or sectioned beef or pork, which is then shaped into roasts, steaks or loaves. The restructuring process consists of three steps: reducing or modifying the particle size, blending, and reforming into the desired shape. Examples of restructured meats are smoked sliced beef and most boneless hams.

INSPECTION PROGRAMS

The Federal Meat Inspection Act of 1906 made inspection mandatory for all meat that was transported across state lines. The Wholesome Meat Act of 1967 required that inspection of meat sold within a state meet requirements at least as stringent as those of the federal system. Federal meat inspection is the responsibility of the Food Safety and Inspection Service, a division of the United States Department of Agriculture (USDA). State inspection is the responsibility of each state government, with partial funding by the federal government.

These federal and state inspection programs assure that only healthy animals are used for meat and guarantee that facilities and equipment meet sanitation standards. The Meat Inspection Program also includes: inspection of meat at various stages of processing; temperature monitoring for both fresh and cooked meat; review of packaging and labels; control and monitoring the use of additives; and control and monitoring of imported meat.

To be designated as kosher ("fit and proper"), meat must be processed under the supervision of persons authorized by the Jewish faith and must meet the standards of the Mosaic and Talmudic laws.

MEAT GRADES

The USDA meat-grading system, established in 1927, sets standards of quality and yield of edible meat that are used in buying and selling most meat. USDA administers the program, but participation is voluntary. There are federal grades for beef, veal, lamb, and pork; however, there are more grades for beef than for other meats due to the greater variance in age and weight of the animals. A shield-shaped purple stamp indicates the USDA quality grade. In addition, individual packers and retailers may use their own brand names to designate different grade of meat, such as Armour "Star," Monfort "Gold," or Safeway "Lean." Federal designations are:

Beef—Prime, Choice, Select, Standard, Commercial, Utility, Cutter, and Canner. The last four grades are seldom sold in retail stores. Beef Carcasses may also be yield graded according to the predicted percentage of edible meat. The range is from one (highest yield) to five (lowest yield).

Veal & Lamb—Prime, Choice, Good, Standard, Utility, and Cull. There is no Standard grade for lamb. Only lamb may be yield-graded from U.S. No. 1 (highest yield) to U.S. No. 5 (lowest yield).

Pork—U.S. No. 1, U.S. No. 2, U.S. No. 3, and U.S. No. 4. The basis for grading pork is the cutability of the carcass and the quality of the meat. The range is from U.S. No. 1 (highest yield) to U.S. No. 4 (lowest yield).

MEAT STORAGE AND USE

Proper storage is essential for maintaining food safety and quality. When meat spoils, the color, odor, and texture deteriorate. When food is contaminated it is invaded by microorganisms, which may not be detected by an off odor or color, but can cause food poisoning and infections in humans. Fortunately, most bacteria that can cause food poisoning cannot grow at low refrigerator temperatures. Recommended storage time for maximum quality of meats at refrigerator and freezer temperatures are shown in Table 4.2. Following are a few guidelines for safe meat storage.

- Fresh meat should be stored in the coldest part of the refrigerator and used within a few days. The temperature should be as low as possible, approximately 36°F to 40°F (2 to 4° C).

- Meat can be kept in the original supermarket wrap in the refrigerator for several days or in the freezer for up to two weeks.

- For longer freezer storage, meat should be tightly rewrapped in moisture-vapor-proof material designed for freezer use (for example, foil, freezer paper, and some transparent wraps and food storage bags). Be sure to label and date the packages and store at 0° F (-18°C) or lower.

- Cured meats can be refrigerated in the original wrapper, but should be used within one or two weeks.

- Unopened canned hams can be refrigerated until ready to use, unless the label indicates otherwise.

- Cooked meats should be wrapped or placed in tightly covered dishes or containers and refrigerated within two hours after serving.

- To defrost frozen meat, keep it wrapped, and let it stand in the refrigerator. Thaw time will depend on size and thickness of the cut — approximately 4 to 7 hours per pound (0.5 kg) for large roasts, 3 to 5 hours per pound (0.5 kg) for a small roast, and 12 to 14 hours for a 1 inch (2.5 cm) thick steak. Thawed meat can be refrozen if the temperature of the meat has not risen above 40°F (4°C).

TABLE 4.2

Refrigerator and Freezer Storage Timetable

Meat	Cut	Refrigerator 36° to 40°F (2–4°C)	Freezer 0°F or Colder (-18°C)
Fresh	Beef cuts	3 to 4 days	6 to 12 months
	Veal Cuts	1 to 2 days	6 to 9 months
	Pork Cuts	2 to 3 days	6 months
	Lamb Cuts	3 to 5 days	6 to 9 months
	Ground Beef, Veal and Lamb	1 to 2 days	3 to 4 months
	Ground Pork	1 to 2 days	1 to 3 months
	Variety Meats	1 to 2 days	3 to 4 months
	Leftover Cooked Meats	3 to 4 days	2 to 3 months
Processed	Luncheon Meat**	3 to 5 days	1 to 2 months
	Sausage, fresh pork**	2 to 3 days	1 to 2 months
	Sausage, smoked**	1 week	1 to 2 months
	Sausage, dry and semi-dry (unsliced)	2 to 3 days	1 to 2 months
	Frankfurters*	3 to 5 days	1 to 2 months
	Bacon**	1 week	1 to 2 months
	Smoked Ham, whole	1 week	1 month
	Ham Slices**	3 to 4 days	1 to 2 months

* Meat in freezer wrapping.

** If vacuum packaged, check manufacturer's date.

COOKING METHODS

Heat affects the physical properties of meat by making it or keeping it tender, flavorful, and safe to eat. Different cooking methods are used to maximize the flavor and eating enjoyment of various cuts of meat. When heat is applied to meat, two changes occur: muscle fibers become tougher and connective tissue becomes more tender. The time required to cook a specific cut of meat by any method depends on the composition of the meat, size, weight, and shape of the cut.

Tender cuts of meat, cooked by dry-heat methods to a relatively low temperature (rare, for example), result in tender juicy products. Less-tender cuts must be cooked for longer periods by moist-heat methods to soften connective tissue, prevent surface drying, and develop flavor. If using a marinade, refrigerate the meat at least six to eight hours in an acidic liquid, such as wine, vinegar, or lemon juice.

Timetables are guides to approximate cooking times required for meat cuts to reach the desired internal temperature, or desired tenderness.

DRY HEAT METHODS

Roasting—This cooking method is recommended for larger cuts of beef, veal, lamb, and pork. Season meat with herbs, spices, or other seasoning before, during, or after roasting. Place meat fat side up on a rack in a shallow roasting pan. In some roasts, rib bones serve as a rack. Insert meat thermometer to the center of the largest muscle. Do not add water or cover.

Roast in a slow oven, 300°F to 325°F (150°C to 160°C), to the desired doneness. No basting is necessary. Remove roast 5°F (3°C) below desired doneness. The temperature of the meat will rise 5°F (3°C) during standing for 15 to 20 minutes, and the roast will be easier to carve. See Roasting Timetable, Table 4.3, on pages 79–80.

Oven Broiling—Broiling is suitable for tender steaks and chops; kabobs; pork ribs; sliced ham; bacon; butterflied lamb leg; and ground meat patties or meatballs. Steaks and chops should be at least 3/4 inch (2 cm) thick, and ham should be at least 1/2 inch (1 cm) thick. Less tender cuts, such as beef flank or top round steak, and veal, lamb, and pork shoulder chops, may also be broiled, if they have been marinated.

Trim the outer edge of fat. Place thinner cuts of meat (3/4 inch to 1 inch or 2 to 2.5 cm thick) on rack of broiler pan 2 to 3 inches (5 to 7.5 cm) from heat. Allow 3 to 6 inches (7.5 to 15 cm) for thicker cuts. Broil until the top side is brown, then turn and brown the other side. To test steaks or chops for doneness, cut a small slit in the center or close to a bone and check the color of the meat. Season, if desired.

Grilling (Barbecuing)—Grilling is actually a method of broiling. Meat can be grilled on a grid or rack over coals or heated ceramic briquettes. While usually done outdoors, grilling can be done in the kitchen with special types of range tops or small appliances.

Start the fire as the manufacturer's instructions indicate. Spread coals in a single layer after they reach the gray ash stage. Low to medium cooking temperatures should be used. To lower the temperature, spread coals farther apart or raise the grid. To make the fire hotter, move the coals closer together and tap off ash. The temperature of the coals can be checked by cautiously holding your hand 4 inches (10 cm) above the coals. Count the number of seconds before the heat forces the removal of your hand. Check the following chart.

Charcoal Temperature for Grilling

Time	Temperature
2 seconds	Hot (high)
3 seconds	Medium-hot
4 seconds	Medium*
5 seconds	Low*
6 to 7 seconds	Very low

* Recommended cooking temperature for meat.

For direct cooking, arrange coals in a single layer directly under the food. Use this method for steaks, ham slices, chops, kabobs, and other quickly cooked foods. For larger roasts and steaks, which require longer cooking at lower temperatures, use indirect cooking. This method cooks food by reflecting heat, similar to the way in which a conventional oven cooks. Arrange an equal number of briquettes on each side of the grate, ignite, and make sure the coals are burning equally on both sides. If necessary, move the coals from one side to the other. Place an aluminum foil drip pan between the coals. Arrange the meat on the grid above the drip pan. Cover with damper open and cook as the recipe directs. Turning the meat during indirect cooking is not necessary.

Baste meat with marinade or sauce throughout cooking time, if desired. If the basting sauce contains a large amount of sugar or other ingredients which burn easily, limit basting to last 15 to 20 minutes. Cooking time can be affected by the type of equipment used, the cooking temperature maintained, the weather, wind velocity, and desired doneness. Discard any marinade not cooked with the meat.

PAN-BROILING

Pan-broiling is a faster, more convenient method than oven broiling for cooking thinner steaks and chops and bacon. Place the meat in a preheated, heavy frying pan or on a griddle. Most meat cuts have enough fat to prevent sticking; however, the cooking surface may be lightly brushed or sprayed with oil when pan-broiling very lean cuts. No oil is needed with a nonstick frying pan. Do not add water or cover.

For cuts thicker than 1/4 inch (0.6 cm), cook at medium heat, turning occasionally. Cuts less than 1/4 inch (0.6 cm) thick should be cooked at medium-high. Turn more than once to assure even cooking. Remove fat as it accumulates, so that meat does not fry. Cook until brown on both sides. Do not overcook. To test doneness, cut a small slit in the center or close to the bone and check the color of the meat.

Pan-frying—In pan-frying, a small amount of fat is added to the frying pan and fat is allowed to accumulate during cooking. This method is suitable for small or thin cuts of lean meat; thin strips; and pounded, scored, or other tenderized cuts that do not need prolonged heating for tenderization.

Brown meat on both sides in small amount of oil. Some cuts will cook in the fat that comes from the meat. Lean cuts, such as cubed steak or cuts, which are floured or breaded, require additional oil on the surface of the pan to prevent sticking. If there is a coating, seasonings may be added to the coating ingredients. Otherwise, the meat may be seasoned after browning.

Do not cover the meat or the meat will lose its crisp texture. Cook at medium heat, turning occasionally to promote uniform cooking.

Stir-frying—Stir-frying is similar to pan-frying except the food is stirred almost continuously. Cooking is done with high heat, using small or thin pieces of meat. Partially freeze meat to facilitate slicing. Slice meat across the grain into thin, uniform slices, strips, or pieces. If desired, marinate while other ingredients are prepared. Cook meat and vegetables separately and then combine. Place meat in a small amount of hot oil in wok or large frying pan. Stir-fry about a half-pound (0.2 kg) at a time. Cook at a high temperature, sliding the spatula under the meat and turning continuously with a scooping motion.

TABLE 4.3
Roasting Timetable for Meat

Cut	Weight		Oven Temperature		Approximate Cooking Time (Minutes Per Pound)		
	Pounds	Kilograms	°F	°C	Rare	Medium	Well Done
Beef							
Rib Roast	4 to 6	2 to 3	300–325	150–160	26 to 32	27 to 30	40 to 42
	6 to 8	3 to 3.6	300–325	150–160	23 to 25	34 to 38	32 to 35
Rib Eye Roast	4 to 6	2 to 3	350	180	18 to 20	20 to 22	22 to 24
Boneless Rump Roast	4 to 6	2 to 3	300–325	150–160		25 to 27	28 to 30
Round Tip Roast	3.5 to 4	1.6 to 2	300–325	150–160	30 to 35	35 to 38	38 to 40
	6 to 8	3 to 4	300–325	150–160	22 to 25	25 to 30	30 to 35
Top Round Roast	4 to 6	2 to 3	300–325	150–160	20 to 25	25 to 28	28 to 30
Tenderloin Roast–							
Half	2 to 3	1 to 1.5	425	220	35 to 45 total time		
Whole	4 to 6	2 to 3	425	220	45 to 60 total time		
Ground Beef Loaf (9" × 5"; 3.6 x 2 cm)	1.5 to 2.5	0.7 to 1	300–325	150–160	60 to 90 total time		
Veal							
Loin Roast							
Boneless	2 to 3	1 to 1.5	300–325	150–160		18 to 20	22 to 24
Bone-In	3 to 4	1.5 to 2	300–325	150–160		34 to 36	38 to 40
Rib Roast	4 to 5	2 to 2.3	300–325	150–160		25 to 27	29 to 31
Crown Roast 12–14 ribs	7.5 to 9.5	3.4 to 4.3	300–325	150–160		19 to 21	21 to 23
Rib Eye Roast	2 to 3	1 to 1.5	300–325	150–160		26 to 28	30 to 33
Rump Roast							
Boneless	2 to 3	1 to 1.5	300–325	150–160		33 to 35	37 to 40
Shoulder Roast–							
Boneless	2.5 to 3	1 to 1.5	300 – 325	150 – 160		31 to 34	34 to 37
Pork							
Loin Roast							
Center–Bone-In	3 to 5	1.5 to 2.3	325	160		20 to 25	26 to 31
Top Loin,							
Double	3 to 4	1.5 to 2	325	160		29 to 34	33 to 38
Single	2 to 4	1 to 2	325	160		23 to 33	30 to 40
Crown Roast	6 to 10	3 to 4.5	325	160			20 to 25

TABLE 4.3 (CONTINUED)
Roasting Timetable For Meat

Cut	Weight		Oven Temperature		Approximate Cooking Time (Minutes Per Pound)		
	Pounds	Kilograms	°F	°C	Rare	Medium	Well Done
Tenderloin Roast, Whole	0.5 to 1	0.2 to 0.5	425	220		27 to 29 total time	30 to 32 total time
Ribs–Backribs			325	160	Cook until tender—90 to 105 total time		
Lamb							
Leg Roast Whole–							
Bone-In	7 to 9	3 to 4	325	160	15 to 20	20 to 25	25 to 30
Boneless	4 to 7	2 to 3	325	160	25 to 30	30 to 35	35 to 40
Shank–Half	3 to 4	1.5 to 2	325	160	30 to 35	40 to 45	45 to 50
Sirloin–Half	3 to 4	1.5 to 2	325	160	25 to 30	35 to 40	45 to 50
Shoulder Roast– Boneless	3.5 to 5	1.6 to 2.3	325	160	30 to 35	35 to 40	40 to 45

Cut	Weight		Oven Temperature		Final Thermometer Reading		Cooking Time (Minutes Per Pound)
	Pounds	Kilograms	°F	°C	°F	°C	
Pork							
Smoked (Fully cooked)							
Whole–Boneless*	8 to 12	4 to 5	300–325	150–160	140	60	13 to 17
Bone-In**	14 to 16	6 to 7	300–325	150–160	140	60	12 to 14
Half–Boneless*	6 to 8	3 to 3.6	300–325	150–160	140	60	17 to 20
Bone-In**	6 to 8	3 to 3.6	300–325	150–160	140	60	14 to 17
Portion–Boneless	3 to 4	1.5 to 2	300–325	150–160	140	60	20 to 23
Canadian Style Bacon (Fully Cooked)	2 to 4	1 to 2	300–350	150–180	140	60	20 to 30

NOTES: Cooking times are based on meat taken directly from the refrigerator.

Smaller roasts require more minutes per pound than larger roasts.

Cooking times are approximate.

* Add 1/2 cup (125 ml) water, cover tightly, and cook as directed.

** Cover tightly and cook as directed.

MOIST HEAT METHODS

Braising—In some areas of the country, "fricassee" and "braising" are interchangeable terms. Two popular examples of braised meats are pot roast and Swiss steak. To braise, first dredge the meat with flour or coat with crumbs, if desired. Then, slowly brown the meat on all sides in a heavy pan in just enough oil to prevent sticking; drain the drippings; and season the meat with salt, herbs, and spices. Browning develops flavor and color. Cuts with sufficient fat do not require additional fat, unless they have been coated with flour or crumbs.

Next, add a small amount of liquid (2 tablespoons to 1/2 cup; 30 ml to 125 ml) such as water, tomato juice, wine, and meat or vegetable stock. Cover tightly and simmer on low heat on top of the range or in a 300°F (150° to 160°C) oven until meat is fork-tender. Check occasionally to make sure liquid does not dry out; add more if necessary. Sauce or gravy may be made from liquid in the pan. Braising times for various cuts of meat are:

Round Steak (3/4" to 1"; 2 to 2.5 cm)	1 to 1 1/2 hours
Beef Chuck and Pot Roasts	1 3/4 to 3 hours
Shortribs	1 1/2 to 2 1/2 hours
Beef Cubes (1 1/2"; 3.7 cm)	1 1/2 to 2 1/2 hours
Veal Shoulder Roast	2 to 2 1/2 hours
Spareribs and Backribs	1 1/2 hours
Pork Cubes (1" to 1 1/4"; 2.5 to 3 cm)	45 to 60 minutes
Lamb Breast	1 1/2 to 2 hours
Lamb Shanks	1 to 1 1/2 hours

In general, the longer the meat is braised, the more tender it will be. Meat can also be braised in cooking bags designed specifically for oven use. The benefit is a reduction in cooking time for larger cuts of meat. No additional liquid is needed, as moisture is drawn from the meat as it cooks. Follow manufacturer's instructions for best results.

Cooking in liquid—Less tender cuts of meat can be covered with liquid and gently simmered until tender. The temperature of the liquid should not rise about 195°F (90°C), since boiling (212°F or 100°C) toughens meat protein. The three ways to cook in liquid are simmering, stewing, and poaching. Simmering (for large cuts) and stewing (for small pieces) are used for less tender cuts of meats.

To simmer or stew, brown meat on all sides in a large, heavy pan or Dutch oven; drain drippings. Corned beef and cured and smoked pork need not be browned. Completely cover the meat with water or stock and season as desired. Cover tightly and simmer (do not boil) until meat is fork-tender.

When vegetables are to be cooked with the meat, add them whole or in pieces near end of cooking period. When done, remove meat and vegetables to a platter or casserole and keep hot. The cooking liquid may be reduced or thickened for a sauce or gravy. Approximate cooking times are:

Beef for Stew (1" to 2"; 2.5 to 5 cm)	2 to 3 hours
Fresh or Corned Beef Brisket (2 1/2 to 3 1/2 lb.; 1 to 1.5 kg)	2 1/2 to 3 1/2 hours
Beef Shank Cross cuts (1" to 1 1/2"; 2.5 to 3.7 cm)	2 1/2 to 3 1/2 hours
Boneless Breast of Veal	1 to 1 1/2 hours
Lamb and Pork for Stew (1 to 1 1/2 lb.; 0.5 to 0.7 kg)	1 1/2 to 2 hours
Country-Style Pork Ribs	2 to 2 1/2 hours

INTERNAL TEMPERATURE

The degree of doneness for meat can be accurately determined by measuring internal temperature with a standard meat thermometer. The thermometer should be inserted into the roast surface at a slight angle or through the end of the roast so the tip of the thermometer is in the thickest portion of the cut, but is not resting in fat or against the bone. Following are internal temperatures for preparing meat:

Beef and Lamb		
Rare	140°F	60°C
Medium rare	150°F	65°C
Medium	160°F	70°C
Well done	170°F	80°C
Veal		
Medium	160°F	70°C
Well done	170°F	80°C
Pork (fresh and cured)		
Medium	160°F	70°C
Well done	170°F	80°C
Ham (fully cooked)	140°F	60°C

COOKING FROZEN MEATS

Frozen meat may be defrosted in the original wrapper and cooked according to directions for meat that has not been frozen, or it may be unwrapped and cooked in the frozen state. Additional cooking time is needed when cooking frozen meat. Frozen roasts require 30 to 50 percent more cooking time than roasts started at refrigerator temperature. The additional time needed for cooking steaks and chops varies according to surface area and thickness of the meat and the cooking temperature.

Thick frozen steaks, chops, and meat patties must be broiled farther from the heat than those that have been defrosted. This allows the internal temperature to reach the desired level without excessive surface browning. When pan-broiling frozen steaks or chops, a heated frying pan should be used for defrosting the surface rapidly and browning on all sides. The heat should be reduced after browning and the meat turned often, so it will cook evenly.

COOKING VARIETY MEATS

All variety meats may be simmered or braised. Some, such as calves' liver, and lamb or veal kidneys, may be broiled. Regardless of the method, variety meats are usually cooked until well done.

MICROWAVES AND FOOD SYSTEMS

MEAT

The best cuts of meat for microwave cooking are tender, high quality, boneless, and uniform in shape. Bone tends to cause uneven cooking due to the reflective effect on microwaves. Popular meat choices are ground beef; steaks; chops; small roasts; country style ribs or spareribs; smoked and cured meat; and precooked, skinless sausage.

As a general rule when microwaving meat dishes, use MEDIUM (50 percent) power or MEDIUM-LOW (30 percent) power to assure tender, flavorful, and evenly cooked meat.

Following are some additional tips for promoting even cooking:

- Cut meat into uniform pieces or form ground meat into donut-shaped patties, 1/2" (1.2 cm) thick.

- Arrange uniform meat shapes, such as meatballs, in a circle on microwave safe dish, leaving center empty. For chops, place meatier portion to outside of dish, with the bone toward the center.

- Stir meat strips several times during cooking, moving the less cooked pieces to the outside and the more cooked pieces to the inside.

- Stir soups, stews, and other dishes with high liquid content several times to distribute the heat.

- Turn meat over several times or rotate dish 1/4 to 1/2 turn at intervals.

- Meat dishes should be covered with a lid or heavy-duty plastic wrap, turned back at one corner. The steam created will help food cook more evenly. Be sure plastic wrap does not touch the food.

A standing period of 10 to 20 minutes may be needed for heat to penetrate the center of the meat after microwaving. When estimating cooking time, allow for a rise in internal temperature of 5°F to 10°F (3 to 6°C) during standing. The slower the cooking and the smaller the mass, the shorter the standing time required. The following internal temperatures are suggested as a guide for determining microwave cooking:

Beef

Rare	130°F to 135°F	55°C to 57°C
Medium	150°F to 155°F	66°C to 69°C
Well done	160°F to 165°F	71°C to 74°C

Lamb

Medium	150°F to 155°F	66°C to 69°C
Well done	160°F to 165°F	71°C to 74°C

Pork

Well done	160°F to 165°F	71°C to 74°C
Ham (fully cooked)	130°F to 135°F	55°C to 57°C

Fish and Shellfish

NUTRITIVE VALUE

The consumption of fish (finfish and shellfish) has risen dramatically within the last decade, primarily because consumers are increasingly aware that fish is one of the most nutritious foods.

PROTEIN CONTENT

Fish contains a complete protein with little connective tissue that is easily cooked until tender. Most species have 17 to 23 percent protein.

VITAMIN AND MINERAL CONTENT

Fishery products are a good source of many vitamins, especially niacin, B_6 and B_{12}. Fish, particularly shellfish, is known for its mineral content, with certain species having substantial amounts of calcium, iron, copper, chromium and/or zinc and other trace minerals. The flesh of both saltwater and freshwater fish is quite low in sodium content, with finfish averaging about 50 mg of sodium in a 3.5 ounces (100 g) raw portion. Shellfish are higher in sodium. Shrimp and oysters contain 235 and 250 mg of sodium per 3.5 ounces (100 g), respectively. Pickled, canned and smoked fish and surimi seafood have added sodium.

FAT CONTENT

The fat content of fish can range from as low as 1.6 g in ocean perch to 8.6 g in salmon per 3.5 ounces (100 g) portion of uncooked seafood and averages about two grams of fat and 85 kilocalories per 3.5 ounces (100 g).

The lipid in fish is high in polyunsaturated fat. Many fish contain omega-3 fatty acids, which are particularly nutritious forms of fat. The cholesterol content of 3.5 ounces (100 g) portion of finfish averages about 60 mg, similar to amounts found in pork, beef and the dark meat of chicken. Most shellfish have moderate levels of cholesterol ranging from 35 to 90 mg per 3.5 ounces (100 g) portion. Shrimp and squid are higher in cholesterol (about 150 and 200 mg, respectively) but, like all fish, both are low in total fat and saturated fat. Fish is classi-

fied as either "lower fat" or "higher fat" which affects the preferred cookery method. Species with less than five percent fat content are considered "lean" and are usually characterized by a mild flavor and lighter colored flesh. Species that have a fat content of five percent or higher are considered "fat" with a more distinctive flavor and a darker color to the flesh.

MARKET FORMS

There are over 500 species of fish and shellfish commercially available in the United States today. They are marketed fresh, frozen, canned, dried, salted, smoked and in many convenience forms.

Fresh and frozen fish are marketed in various forms. The following are the most common:

Whole or **round** fish are just as they came from the water. Before cooking they must be eviscerated.

Drawn fish have been eviscerated and, usually, scaled.

Dressed fish have been scaled and eviscerated and the head, tail and fins have been removed. A small dressed fish is called **pan-dressed**.

Fillets are the sides of a fish cut away from the backbone. They may be purchased with the skin on or off and may have small rib or pin bones.

Butterfly or **kited fillets** are both sides of a fish cut away from the backbone and left connected by the uncut flesh and skin of the back or belly.

Steaks are cross sections of a large dressed fish, typically $3/4$ to 1 inch (2 to 3 cm) thick. Usually, the skin is left on, and a cross-section of the backbone remains. In the case of very large fish (such as some tuna), the fish is cut into loins before steaking, and therefore has neither skin or bone.

Chunks are cross-section portions of large dressed fish, having a cross-section of backbone as the only bone. They are cut much like a beef or pork roast and are ready to cook as purchased.

Raw breaded fish portions are cut from frozen fish blocks coated with a batter, breaded, packaged and frozen. They are ready to cook as purchased.

Fried fish portions and **fried fish sticks** are cut from frozen fish blocks, coated with a batter, breaded, partially cooked, packaged and frozen. They are ready to heat and serve as purchased.

Canned fish have been dressed or filleted, mechanically sized and placed into a can that is then sealed and heated. Many species of fish are available canned. Among them are salmon, tuna, sardines, mackerel, shrimp, oysters and clams. They may be served as purchased.

Salmon, canned on the Pacific Coast, is usually sold by the name of the species. These differ in fat content, color, texture and flavor. The higher priced varieties are deeper red in color and have a higher oil content than the less costly ones (in descending order: sockeye; chinook or king; coho or silver; and chum or keta).

Tuna canned in the United States is produced from several species of tuna. Albacore, a very light meat, is the only tuna to be labeled white-meat tuna. The other species (yellowfin, blackfin, bluefin and skipjack) are labeled light-meat tuna. Fancy or solid pack tuna usually contains three or four large pieces packed in oil or water. Chunk, flaked and grated-style packs contain mechanically sized pieces packed in oil or water.

Sardines are small, immature sea herrings that are packed in water, oil, mustard or tomato sauce.

Cured fish are either salted or smoked and include salt herring, salmon, salmon eggs, chubs, sablefish, sturgeon, whitefish, mackerel and mullet. Lox is a mildly cured salmon.

Surimi seafood is an imitation shellfish product made from a mild-flavored finfish, usually Alaskan pollock. The freshly caught fish is made into a paste and them formed into the shape, texture and color of a shellfish, most commonly crab meat. The processing involves the addition of certain stabilizers, binders and flavorings, often including real shellfish or a shellfish extract. Surimi seafood is pre-cooked and may or may not be breaded. The nutrient value of unbreaded surimi seafood is similar to that of finfish, with the exception of a higher sodium content. It is generally lower in cost than the shellfish it imitates.

SIGNS OF FRESHNESS

FRESH FINFISH

Good quality fresh (unfrozen) finfish have the following characteristics:

- The flesh is firm, not separating or soft. It springs back when pressed gently.
- The odor is fresh and mild. Do a "nasal appraisal" before purchasing.
- The eyes are bright, clear and often protruding. As the fish loses freshness, the eyes become cloudy, pink and sunken.
- The gills are bright red or pink and are free from slime. Avoid fish with dull colored gills that are gray, brown or green.
- The skin is shiny, with scales that adhere tightly. Characteristic colors and markings may begin to fade soon after the fish is removed from the water.
- The intestinal cavity is clean and pink.

FRESH SHELLFISH

Different species of shellfish vary greatly in terms of market forms and physical characteristics. Look for the following signs of freshness when buying fresh (unfrozen) shellfish:

- The odor is fresh and mild.
- The flesh of shrimp and spiny (southern) lobster is firm. The shell is free of black spots, called melanosis, which are a sign of aging.
- Oysters, clams and muscles in the shell are alive, as indicated by a tightly closed shell or one that closes upon handling. If a shell opens slightly, tap it with a knife, and it will close. Discard any that do not.
- Uncooked crabs and northern lobsters are alive, as indicated by leg movement. They will not be very active if they have been refrigerated, but they should move a least a little. Also, the tail of a live lobster curls under tightly when it is handled. Discard any uncooked crabs and northern lobsters that do not show these signs of life.

- Fresh shucked oysters are in a clear, slightly milky or light gray liquid, called liquor. Oysters are usually creamy white, but there are color variations depending on variety and diet.

- The smaller bay and calico scallops are usually creamy white, though there may be some normal light tan or pink coloration. The larger sea scallops are generally creamy white, though they may show some normal light orange or pink color.

- Crab meat has a fresh, mild odor.

FROZEN FISH

Fishery products that are sold in the frozen form are usually packed during the seasons of abundance, at the peak of freshness, and are held in cold storage until ready for distribution. Thus, the consumer is given the opportunity to select varied species of fish throughout the year and at better prices. High quality frozen fish that are properly processed, packaged and held will remain in good condition for relatively long periods of time.

Frozen fish of good quality have the following characteristics:

- The fish is solidly frozen when purchased.

- The flesh has no discoloration or freezer burn (a dry, white or grayish "cottony" appearance.

- The fish has little or no odor. When thawed, it has a mild odor.

- The wrapping is intact and is of moisture-vapor-proof material.

- There is little or no air space between the fish and the wrapping, and there are no ice crystals or frost in the package which indicate that a variation in temperature has occurred.

HOW MUCH TO BUY

The amount of fish to buy per serving varies with the market form, the recipe to be used and the yield of edible flesh for the species. Consider about 3 ounces (85 g) of cooked, edible flesh as a serving. The following list can help you determine how much fish to buy per serving.

Finfish

Whole	$^3/_4$ pound (350 g)
Dressed, pan-dressed	$^1/_2$ pound (250 g)
Fillets, steaks	$^1/_3$ pound (140 g)
Portions	$^1/_3$ pound (140 g)
Sticks	$^1/_4$ pound (125 g)
Canned	$^1/_6$ pound (75 g)

Shellfish

Crab, cooked meat	$^1/_4$ pound (125 g)
Crabs, live	1 to 2 pounds (500 g to 1 kg)
Lobster, cooked meat	$^1/_3$ pound (140 g)
Lobster, whole	1 to 2 pounds (500 g to 1 kg)
Mussels, live	12
Oysters, clams, live	6
Oysters, clams, shucked	$^1/_4$ pint (125 ml)
Oysters, clams, breaded	4 to 6
Scallops, shucked	$^1/_4$ pound (125 g)
Shrimp, unpeeled, headless	$^1/_2$ pound (250 g)
Shrimp, cooked, peeled	$^1/_4$ pound (125 g)
Shrimp, breaded	4 to 6
Squid, whole	$^1/_2$ pound (250 g)
Squid, cleaned	$^1/_4$ pound (125 g)

STORING AND THAWING

Fresh fishery products that are to be cooked the day of purchase should be stored immediately in the coldest part of the refrigerator in the original wrapper. Fish that is to be held up to two days before cooking should be rinsed, patted dry and rewrapped in clean wrapping before refrigerating. For optimum quality, do not hold fresh fish in the refrigerator longer than two days before cooking. A storage temperature of 32° to 40°F (4°C) is needed to maintain the quality of the product.

Fish that is purchased fresh may be successfully frozen at home. While there are a number of ways to properly freeze fish, the primary concern with all methods is to keep air away from the product. It should be wrapped in a moisture-vapor-proof wrapping (paper, plastic and/or foil) or frozen in water.

Frozen fishery products should be placed in the freezer in their original moisture-vapor-proof wrapping immediately after purchase, unless the fish is to be thawed for cooking. A storage temperature of 0°F (-18°C) or lower is necessary to maintain quality. At higher temperatures, chemical changes cause fish to lose color, flavor, texture, and nutritive value. Generally, higher fat species may be held in the freezer up to four months without significant loss or quality, and lean species may be frozen for up to one year. It is, however, best to use frozen products soon after purchase for optimal quality.

When thawing fish, take into consideration the following:

- Schedule thawing so that the fish will be cooked soon after it is thawed. Do not hold thawed fish longer than a day before cooking.

- Place frozen packages in the refrigerator to thaw, allowing 18 to 24 hours or longer for thawing a one pound (500 g) package.

- If quicker thawing of unbreaded product is necessary, place the frozen package under cold running water or in cold water to thaw. Allow 1/2 to 1 hour per pound (500 g) of frozen product.

- Thaw fish in the microwave on the defrost setting for four to five minutes per pound (500 g), just until it is pliable enough to handle.

- Do not thaw fish at room temperature or in warm water.

- Avoid refreezing fish, as it results in a lower quality product with a higher bacteria count.

- For prepared products including fish sticks and portions, follow the manufacturer's instructions.

- Frozen fillets and steaks may be cooked without thawing if additional cooking time is allowed.

SAFE HANDLING

All raw foods, including finfish and shellfish, contain bacteria. Handle fish as you would any perishable food product: keep it properly refrigerated, cook it adequately, and handle it with safety in mind. To maintain wholesomeness and quality, always follow these guidelines:

- Know your fish retailer. Buy fishery products from approved, licensed stores and markets.

- Purchase raw shellfish carefully. Buy raw oysters, clams and mussels only from approved, licensed, reputable sources. If in doubt, ask to see the certified shipper's tag that accompanies all live mollusks; look for a shipper number on containers of shucked oysters.

- Keep fish cold.

- Refrigerate live shellfish properly. Live oysters, clams and mussels should be stored under well-ventilated refrigeration, not in airtight plastic bags or storage containers. Burlap, net or heavy paper bags work well, as do plastic bags that are left open. Live lobsters and crabs should be stored similarly, with a covering of damp paper towels.

- Keep live shellfish alive until cooking. Discard any that have died during storage.

- Do not cross-contaminate. Handle and store raw and cooked products separately. Once a cutting board or colander or platter has been in contact with raw finfish or shellfish, wash it thoroughly before using it for cooked products.

- Cook fish thoroughly.

- Harvest fish carefully. If you catch fish or shellfish from local waters, make sure the waters are approved for harvest. Check with your local state Department of Health. Always carry plenty of ice and do not catch more than you can chill immediately.

EATING RAW FISH AND SHELLFISH

Many consumers enjoy raw fish. Just as the consumption of raw or undercooked beef, pork, chicken and eggs carries a degree of risk of foodborne gastrointestinal illness, so does the consumption of undercooked fish, raw fish (i.e., sashimi, sushi, ceviche) and raw oysters, clams and mussels. Here are some tips to reduce the risk:

- Buy only the highest quality fish and shellfish; keep it properly refrigerated, and use it the same day of purchase.

- If not previously frozen, hard freeze fish intended for preparation of sashimi, sushi or ceviche for seven days, to avoid the slight risk associated with parasites.

- Be certain that oysters, clams, and mussels come from certified growing waters. Make certain they are kept well refrigerated.

- Never eat raw or undercooked fishery products if you are seriously ill with a disease that compromises the immune system, such as cancer, AIDS, alcoholism, liver disease or kidney disease.

HOW TO COOK FISH

Fish is delicious if properly cooked. Cooking fish develops its flavor, softens the small amount of connective tissue present and gives the desired texture. Over-cooking fish toughens it, dries it out, destroys the fine flavor and reduces the yield.

Raw fish has a watery, translucent look. During the cooking process the moisture in fish becomes milky colored giving the flesh an opaque, whitish tint. When this occurs at the center part of the fish, the fish is properly cooked.

To estimate how long to cook a fish, use the "10-minute rule" as a guideline, unless the fish is to be microwaved or deep-fried. Follow these steps:

1. Measure the fish at its thickest point. If other ingredients are added, include them in the measurement.

2. Plan to cook the fish for approximately 10 minutes per inch (2.5 cm) of thickness.

3. Before the estimated cooking time is complete, use a fork or knife, check the fish for doneness at the thickest point. Remove the fish from the heat when finfish are flaky, and shellfish are tender; both should no longer be translucent in the center. An internal temperature of 145°F (60°C) ensures the fish is safe to eat. The cooking time required will vary somewhat according to the temperature and cooking method used. Check for doneness early and often.

4. Double the estimated cooking for frozen fish that has not been thawed.

Most fish are quite delicate and tend to break up easily, so handle as little and as gently as possible during and after cooking to preserve appearance.

When choosing a cooking method and recipe for fish, carefully consider the species to be used. Fat fish are usually distinctive in flavor and can therefore be used in recipes with other robust flavors; the higher fat content also helps maintain moisture in the cooked product. Lean species of fish and shellfish should be used with ingredients that will not overpower their milder flavors, and care should be taken to prevent the product from drying out.

One may substitute a different species for the one specified in a recipe as long as it has similar characteristics. Consider the texture, flavor and fat content.

COOKING METHODS

BAKING

Whole fish lend themselves well to baking, as do fillets and steaks. Cut fish into serving-size portions. Arrange on a well-oiled baking dish, and baste with melted butter or margarine, oil or a marinade. If desired, sprinkle with herbs or spices, citrus juice, or wine. Bake at 375°F (190°C). Turning is not necessary.

BROILING

Choose pan-dressed fish, fillets, steaks or shellfish for broiling. Prepare the fish as for baking. Place under the broiler, about 3 or 4 inches (8 to 10 cm) from the source of heat. Place thicker cuts of fish farther from the heat than thin ones. Baste frequently; turn halfway through cooking time unless fish is very thin.

DEEP-FAT FRYING

Choose fillets, shucked shellfish and small dressed fish for this method. In a deep-fat fryer, heat enough vegetable oil to float the fish but do not fill the fryer more than half full. Heat the oil to 375°F (190°C). Cut the fish into serving-size portions. Dip the fish into seasoned milk or beaten egg and then into seasoned crumbs, cornmeal or flour. Or coat with flour and then dip into a batter. Arrange in a fryer basket to be lowered into the fryer or drop gently into the oil and fry until golden brown, flaky and no longer translucent in the center. Drain on absorbent paper.

PAN-FRYING

Pan-frying is an excellent way to cook pan-dressed fish, fillets and steaks. Prepare the fish as for deep-fat frying. Heat 1/4 to 1/2 inch (6 mm to 1 cm) oil or other fat in a heavy fry pan. Place one layer of fish in the hot fat, taking care not to overload the pan and thus cool the fat. Fry until golden brown on one side, then turn and brown on the other side. Drain on absorbent paper.

OVEN-FRYING

Fillets are the preferred market form to use for oven-frying. Cut the fish into serving-size portions. Dip the pieces in seasoned milk or other liquid and coat with flour or crumbs such as crackers, bread or a snack food. Place the fish on a shallow, well-oiled baking pan drizzle lightly with oil and bake at 450°F (240°C). Turning is not necessary.

POACHING

Any cut of fish may be poached. Often, shellfish is referred to as "boiled," when actually it is prepared by poaching or simmering. Place just enough liquid to cover the finfish into a shallow, wide pan, such as a large fry pan. A Dutch oven or stockpot may be used for shellfish. The liquid may be plain or lightly salted water, water seasoned with herbs and spices, milk, or a mixture of white wine and water. Cover the pan and heat until the liquid simmers. Gently add the fish, cover, and simmer. After the fish is removed, the liquid may then be reduced and thickened to make a sauce for the fish.

STEAMING

Choose any market form of fish for steaming. A steam cooker or any deep pot with a tight cover may be used for steaming fish. If a steaming rack or basket is not available, use a plate or cake rack placed atop an inverted cup or bowl to hold the fish above the water. Pour about 2 inches (5 cm) of water into the pot and bring to a boil. The water may be plain or seasoned with herbs, spices or wine. Place the fish on a lightly oiled rack, basket or plate over the boiling water. Cover the pot tightly and steam.

CHARCOAL BROILING

Fillets, steaks and shellfish are good choices for this cooking method. Before heating the grill, lightly brush the grill and the fish with oil. For fish that is 1 inch (3 cm) thick or less, place the grill 2 to 4 inches (5 to 10 cm) away from the coals; for thicker pieces, 5 to 6 inches (12 to 15 cm) away. For particularly delicate fish, use a hinged, wire fish or hamburger basket (oiled); wooden or metal skewers may be used for shrimp or scallops. Turn the fish halfway through the estimated cooking time, basting frequently with oil or sauce.

MICROWAVE COOKING

Small whole fish, fillets and steaks are particularly good choices for microwave cookery. To achieve the best results when microwaving fish, special consideration should be given to the quantity and shape of the fish, the size, shape of the dish, and the wattage and power settings of the oven. Because there are no standard power settings among most oven brands, the cooking time suggested in recipes should be considered only a guideline. Frequent checking for doneness is required, and notations should be made on actual cooking time for future reference. Use the following techniques when microwaving fish:

- Use a shallow dish to allow the fish maximum exposure to the microwaves. Cover the dish with a paper towel, waxed paper or plastic wrap (vented).
- Arrange the fillets with the thicker parts toward the outside of the dish. The thinner parts, separated by plastic wrap, may be overlapped in the center.
- Use the high power setting and plan to microwave the fish 3 to 5 minutes per pound (500 g). Rotate the dish halfway through the estimated cooking time; turning the fish is not necessary. Check the fish early and often for doneness. If needed, continue to microwave and check for doneness at one-minute intervals.

COOKING SHELLFISH

There are special considerations for cooking each species of shellfish, particularly regarding cooking times. Cook shellfish lightly, looking for the following signs of doneness.

- When properly cooked, shrimp turn pink with an opaque rather than translucent center. One pound (500 g) of medium or large shrimp require only 3 to 5 minutes of cooking when being boiled or steamed.

- Shucked mollusks, such as oysters, clams and mussels, become plump and opaque when cooked. The edges of oysters start to curl.

- Mollusks in the shell open when they are done. Remove individual pieces from the heat as they open. Discard any that do not open.

- Scallops turn milky white in the center when done. Depending on the size and the cooking method, scallops take from 2 to 5 minutes to cook.

- "Boiled" northern lobster turns bright red. When properly cooked, the center of the meat is no longer translucent. Allow 5 to 6 minutes per pound (500 g) in simmering water.

- Spiny lobster turns bright orange with an opaque rather than translucent center. Cooking time in simmering water varies from 8 to 15 minutes, depending on the size and market form.

AQUACULTURE

Aquaculture is the raising of fish or shellfish in tanks, ponds, waterways, or net pens and is one of the fastest growing sectors of the U.S. agricultural industry.

- Worldwide consumer demand for fish continues to increase, while harvests of wild stocks remain static.

- Cultured fish provide consumers with fresh fish year round.

- The aquaculture industry is developing many species, such as hybrid striped bass and tilapia.

Among the species increasingly cultured are catfish, crawfish, shrimp, salmon, tilapia, oysters, clams, scallops, and lobster.

INSPECTION AND GRADING

By the authority of the federal Food, Drug, and Cosmetic Safety Act, the Food and Drug Administration (FDA) of the U.S. Department of Health and Human Services is primarily responsible for the regulation of fish and seafood. FDA conducts inspections of seafood processing operations and evaluates fish handling procedures within each facility. Inspectors analyze and test the products produced in these plants for filth, decomposition and contaminants. Imported seafood is also overseen by FDA, which is responsible for wharf examination and product testing.

The National Marine Fisheries Service of the U.S. Department of Commerce (USDC), the Environmental Protection Agency and coastal state governments all participate in seafood regulation programs which collectively and comprehensively monitor seafood safety.

Inspection and grading of fish and fishery products are voluntary services that the U.S. Department of Commerce makes available to the seafood industry on a fee-for-service basis. Labels of inspected and/or graded fishery products that meet federal standards then may carry U.S. grade and inspection marks with a grade of the product and a statement that the product was packed under federal inspection.

Quality grades include Grade A fishery products, which are top or best quality, uniform in size, practically free from blemishes or defects and of good flavor; Grade B, which is good quality but not as uniform in size or free from blemishes and defects as the top grade; and Grade C, which is fairly good quality, considered wholesome and nutritious, but may not be as attractive in appearance as the other grades.

USDC has issued standards for many frozen precooked and raw breaded fishery products to establish minimum percentages of flesh content. The following table (Table 5.1) lists the minimum percent flesh for breaded fishery products bearing the USDC Grade A inspection marks.

TABLE 5.1

Minimum Flesh Content Requirements for USDC-Inspected Products

Product	USDC Grade Marks (Min. % of Flesh by Weight)	PUFI Mark[1] (Min. % of Flesh by Weight)
Fish		
Raw Breaded Fillets	—	50%[2]
Precooked Breaded Fillets	—	50%
Precooked Crispy/Crunchy Fillets	—	50%
Precooked Battered Fillets	—	40%
Raw Breaded Portions	75%	50%
Precooked Breaded Portions	65%	50%
Precooked Battered Portions	—	40%
Raw Breaded Sticks	72%	50%
Precooked Breaded Sticks	60%	50%
Precooked Battered Sticks	—	40%
Scallops		
Raw Breaded Scallops	50%	50%
Precooked Breaded Scallops	50%	50%
Precooked Crispy/Crunchy Scallops	—	50%
Precooked Battered Scallops	—	40%
Shrimp		
Lightly Breaded Shrimp	65%	65%[3]
Raw Breaded Shrimp	50%	50%[3]
Precooked Crispy/Crunchy Shrimp	—	50%
Precooked Battered Shrimp	—	40%
Precooked Breaded Shrimp	—	No minimum. Encouraged to put percent on label.[4]
Oysters		
Raw Breaded Oysters	—	50%[5]
Precooked Breaded Oysters	—	50%[5]
Precooked Crispy/Crunchy Oysters	—	50%[5]
Precooked Battered Oysters	—	40%[5]
Miscellaneous		
Fish and Seafood Cakes	—	35%
Extruded and Breaded Products	—	35%

[1] PUFI: nonstandardized breaded and battered products. (PUFI = Packed Under Federal Inspection)

[2] No USDC grading standard currently exists.

[3] FDA Standard of Identity requires any product to have 50 percent shrimp flesh by weight. If a product is labeled "lightly" breaded it must contain 65 percent shrimp flesh.

[4] Any product with a Standard of Identity which contains less flesh than the standard requires must be labeled imitation.

[5] Flesh content on oyster products can only be determined on an input weight basis during production.

Poultry and Eggs

POULTRY

Poultry is marketed as ready-to-cook raw poultry and ready-to-eat forms. More poultry is sold as parts, often skinless and boneless, and as further processed products than as whole birds. Many further processed food products include poultry meat as a main ingredient. New packaging techniques, such as modified atmosphere packaging, shelf-stable products and sous vide products (dishes with fresh ingredients which are vacuum-packed, cooked under vacuum and chilled), have expanded the variety of refrigerated, canned, and frozen items available.

FEDERAL POULTRY PROGRAMS

All poultry slaughtered in the United States is inspected for wholesomeness by the federal government (USDA) or by an equivalent state system. Federal inspection is mandatory for slaughtered poultry moving in interstate commerce. Poultry moving in intrastate commerce is subject to state inspection.

Birds are examined inside and out by an official inspector. Poultry may also be graded for quality by an official grader through a voluntary program. Officially inspected and graded poultry is prepared in processing plants that meet USDA sanitary requirements. The inspection mark (a circle stating "Inspected for Wholesomeness by the U.S. Department of Agriculture") and the grade mark (a shield stating USDA Grade A, B, or C) appear on the package label.

Grades for poultry have been established by USDA. Most poultry identified with a grade mark will be Grade A. It will be fully fleshed, meaty, well formed, have a uniform fat covering, a good clean appearance, and be free from defects such as skin cuts and tears, pinfeathers, bruises, and missing or broken bones.

For each kind of poultry, USDA has also established classes, essentially based on the age of the bird. Unless specified, the birds may be of either sex. The age of the bird affects its tenderness and dictates the cooking method to use. Young birds are tender-meated and can be cooked by any method but are best roasted, broiled, or fried. Mature birds are less tender and need to be braised or stewed.

SAFE HANDLING

Poultry, like many other foods, may contain salmonella bacteria, but proper cooking and handling methods can eliminate any threat to health. Thorough washing under cool running water prior to cooking greatly reduces bacterial hazards. Since salmonella are easily destroyed by heat, all poultry should be cooked to the well-done stage, 170°F (80°C). Juices from the cooked meat should be clear, not pink.

Always thaw poultry in the refrigerator or in cold water, **not** on the counter. Bacteria multiply rapidly at room temperature.

Wash hands, counter tops and utensils in hot soapy water between each step in poultry preparation. Bacteria present on raw meat or poultry can get into other food if exposed to the same utensils. For instance, do not cut up raw poultry and then use the same knife or cutting board to prepare other foods unless the knife and board have been washed thoroughly. Be sure the platter that carries the cooked meat to the table is not the same platter that carried the raw meat to the grill.

Cooked poultry that is not eaten immediately should be kept either hot, 140°F (60°C) or hotter, **or** cold, 40°F (4°C) or colder. Thoroughly reheat leftovers before eating. Bring gravies to a rolling boil before serving.

Store poultry for picnics or lunches in an insulated container until ready to eat.

In the Refrigerator

Store chilled raw poultry in its transparent wrap for one or two days in the coldest part of the refrigerator. If wrapped only in market paper, unwrap, place on a tray, cover with waxed paper and refrigerate. Wrap and store giblets separately.

Cool cooked poultry quickly by wrapping loosely and

TABLE 6.1
Freezer Storage Timetable

Poultry Form	Months
Uncooked poultry	
Chicken	
Cut-up	9
Giblets	3
Whole	12
Duck, whole	6
Goose, whole	6
Turkey	
Cut-up	6
Whole	12
Cooked poultry (slices or pieces)	
Covered with broth	6
Not covered with broth	1
Cooked poultry dishes	4 to 6
Fried chicken	4
Poultry gravy or broth	2 to 3

TABLE 6.2
Thawing Timetable in Refrigerator for Poultry

Poultry Type and Size	Time Required
Chickens	
4 pounds (2 kg) or more	1 to 1½ days
Less than 4 pounds (2 kg)	12 to 16 hours
Ducks	
3 to 7 pounds (1 to 3 kg)	1 to 1½ days
Geese	
6 to 12 pounds (3 to 5 kg)	1 to 2 days
Turkeys	
4 to 12 pounds (2 to 5 kg)	1 to 2 days
12 to 20 pounds (5 to 9 kg)	2 to 3 days
20 to 24 pounds (9 to 10 kg)	3 to 4 days
Pieces of large turkey (half, quarter, half breast)	1 to 2 days
Cut-up pieces	3 to 9 hours
Boneless roasts	12 to 18 hours

setting it in the coolest part of the refrigerator.

Remove stuffing from cooked stuffed birds and store in a separate covered container in the refrigerator. Refrigerate broth or gravy promptly.

Use cooked poultry within three to four days. Use stuffing, broth or gravy within one or two days.

In the Freezer

Both uncooked and cooked poultry may be frozen and stored for several months at 0°F (18°C) or colder. (See Table 6.1 for recommended storage times for maximum quality). Do not stuff uncooked poultry before freezing. Remove stuffing from cooked poultry and freeze meat and stuffing separately.

Thawing

Keep poultry frozen until ready to thaw or cook. Thaw in the refrigerator in original wrapping (see Table 6.2) **or** in cold water, changing the water every 30 minutes **or** in the microwave following manufacturer's directions.

Cook promptly after thawing. Do **not** thaw poultry at room temperature.

COOKING METHODS

All poultry should be washed in cold water before cooking and cooked to the well-done stage. Juices should run clear. Whole birds or parts may be roasted, fried, braised or stewed.

DRY HEAT METHODS

Roasting

Allowing about 1 cup (250 ml) stuffing per pound (0.5 kg) of bird, stuff poultry lightly just before roasting. Do **not** stuff the day before.

Roast on rack in uncovered shallow pan. Do not add water.

Do **not** partly roast poultry one day and complete roasting the next day.

Roast stuffed or unstuffed poultry in preheated 325°F

(160°C) oven. Do **not** cook at low temperatures overnight.

You can use a meat thermometer to determine doneness of medium to large birds. Temperature in the inner thigh (not touching the bone) should reach 180° to 185° F (80° to 85°C) or for turkey roasts, 170° F (77°C). Stuffing should reach 165°F (75°C). For approximate total roasting times of various poultry products see Table 6.3.

To roast commercially stuffed poultry, follow directions on the package. Frozen commercially stuffed poultry is ready for the oven and must not be thawed before cooking.

Broiling

Chicken, small fryer-roaster turkeys and ducklings can be broiled satisfactorily. Cut into halves, quarters or pieces. Place cut chicken pieces on rack of broiler pan 5 to 6 inches (7.5 to 15 cm) from heat. Broil on one side until browned, turn and broil until done. Chicken will require 20 to 30 minutes on the first side and 15 to 20 minutes after turning. Turkey and duck pieces, being thicker, require longer broiler time. Allow 60 to 75 minutes total.

TABLE 6.3
Roasting Time for Poultry

	Purchased Weight[1]		Approximate Total Roasting Time at 325°F (160°C)	
			Fresh or thawed poultry 32° to 40°F, 4°C	Frozen poultry 0°F, -18°C or below
Poultry form	Pounds	Kilograms	(Hours)	(Hours)
Chickens				
Whole				
Broiler–Fryer	$2^{1}/_4 – 3^{1}/_4$	$1 – 1.5$	$1^{3}/_4 – 2^{1}/_2$	$2 – 3$
Roaster	$3^{1}/_4 – 4^{1}/_2$	$1.5 – 2$	$2 – 3$	$2^{1}/_2 – 3^{1}/_2$
Capon	$5 – 8^{1}/_2$	$2 – 4$	$3 – 4$	
Rock Cornish game hen	$1 – 2$	$0.5 – 1$	$1^{1}/_2 – 2$	
Pieces	$^{1}/_4 – ^{3}/_4$	$0.1 – 0.3$	$1 – 2^{1}/_2$	$1 – 2^{3}/_4$
Ducks	$4 – 6$	$2 – 3$	$2^{1}/_2 – 4$	$3^{1}/_2 – 4^{1}/_2$
Geese	$6 – 8$	$3 – 4$	$3 – 3^{1}/_2$	
Turkeys				
Whole, unstuffed	$6 – 8$	$3 – 4$	$3 – 3^{1}/_2$	
	$8 – 12$	$4 – 5$	$3^{1}/_2 – 4^{1}/_2$	
	$12 – 16$	$5 – 7$	$4^{1}/_2 – 5^{1}/_2$	$5^{1}/_2 – 6^{1}/_2$
	$16 – 20$	$7 – 9$	$5^{1}/_2 – 6^{1}/_2$	
	$20 – 24$	$9 – 10$	$6^{1}/_2 – 7$	$7 – 8^{1}/_2$
Halves, whole breast	$5 – 11$	$2 – 5$	$3 – 5^{1}/_2$	$4^{1}/_2 – 6^{1}/_4$
Quarters—thighs, drumsticks	$1 – 3$	$0.5 – 1.5$	$2 – 3^{1}/_2$	$2 – 3^{3}/_4$
Boneless roasts	$3 – 10$	$1.5 – 5$	$3 – 4$	

[1] Weight of giblets and neck included for whole poultry.

Outdoor Broiling

Set broiler racks 6 to 8 inches (15 to 20 cm) from the coals. If meat cooks too fast, raise rack or spread coals farther apart. Broil, turning occasionally, about one hour for chicken, longer for turkey.

Rotisserie Cooking

Whole poultry or large pieces can be cooked on rotisserie equipment. Follow manufacturer's directions.

Frying

Young chickens, capons, Rock Cornish game hens and small turkeys can be fried. Serving size pieces are easiest to handle and may be coated or uncoated.

To avoid excess fat in the diet, it is wise to prepare fried poultry with a minimum of fat. Removing skin before cooking also reduces fat.

To pan-fry, heat enough fat or oil, about 2 or 3 table-spoons (30 to 45 ml), to cover the bottom of a heavy frying pan. Place meatier pieces in the center of pan, smaller pieces at edges. Brown over medium heat on one side, turn and brown other side. For chicken, continue to cook slowly, uncovered, until tender, 30 to 45 minutes, or cook in 350°F (180°C) oven until done. For turkey, cover tightly after browning and cook slowly 45 to 60 minutes or until done, turning occasionally. Uncover for the last few minutes to re-crisp, if desired.

To oven-fry, place poultry pieces in a well-greased baking pan. Cook, skin side down, in 400°F (200°C) oven for 30 minutes. Turn and cook 20 to 30 minutes longer or until done. Turkey pieces will need a longer cooking time.

Deep-fat frying, while it produces a desirable crisp coating, also results in a cooked product very high in fat content. To cook by this method, coat poultry pieces with a thin batter, flour, or crumbs. Use a deep kettle, filled no

TABLE 6.4
Braising Timetable for Poultry

| Poultry form | Ready-to-Cook Weight[1] | | Approximate Total Braising Time | |
	Pounds	Kilograms	Fresh or thawed poultry 32° to 40°F, 4°C (Hours)	Frozen poultry 0°F, -18°C or below (Hours)
Chicken				
Whole				
Broiler–Fryer	2¼ – 3¼	1 – 1.5	1½ – 2¼	2 – 2½
Roaster	3¼ – 4½	1.5 – 2	1¾ – 2½	2½ – 3½
Pieces	¼ – ¾	0.1 – 0.3	½ – 1	¾ – 1½
Wings	¼ – less	0.1	1	1¼
Turkey				
Whole	12 – 16	5 – 7	3½ – 5	4½ – 6
	16 – 20	7 – 9	4½ – 6½	5½ – 7½
	20 – 24	9 – 10	6 – 8	7 – 9
Half, breast	5 – 11	2 – 5	2 – 3¼	3 – 4½
Quarters—thighs, drumsticks	1 – 3	0.5 – 1.5	1 – 2	1¼ – 2¼
Wings	¾ – 1¼	0.3 – 0.5	2 – 2½	2 – 2½
Boneless roast	3 – 10	1.5 – 5	1½ – 5	

[1] Weight of giblets and neck included for whole poultry.

more than half full with melted fat or oil (deep enough to cover chicken pieces). Heat fat to 365°F (180°C). Check temperature with a deep-fat thermometer. Fry a few pieces at a time for 10 to 15 minutes. Drain on paper towels to remove excess fat.

Smoking

In some areas, smoking is a popular method for cooking turkeys. Smoking may be done in equipment designed for the purpose or in a covered backyard grill, following manufacturer's directions.

Prepare thawed or fresh unstuffed turkey as for roasting. Soaking in a brine solution before cooking lends a characteristic flavor but is not essential.

Allow the bed of coals to become hot. Add water to the drip pan under the turkey. Temperature of the smoker should be kept between 225°F and 300°F (107°C and 150°C) throughout the cooking. To maintain this temperature, add coals every one to two hours. Cook until meat thermometer registers 180°F (80°C) in thigh and 170°F (75°C) in the breast meat, and juices run clear. Cooking time will vary depending on outside temperature, wind, and altitude and may take as long as eight hours. Smoking methods may also be used for preparing geese, ducks and pheasants.

MOIST HEAT METHODS

Braising

Mature, less tender poultry often requires braising (cooking slowly in a tightly covered pan with a small amount of moisture). Braising of young, tender poultry and boneless turkey roasts is faster than roasting. (See Table 6.4 for braising timetable.)

Place whole or cut up poultry on a rack in roaster or heavy pan and cover tightly. Cook in 400°F (200°C) oven until leg joints move easily and flesh on leg is soft and pliable. To brown, uncover for last 30 minutes of cooking of whole large turkeys or a shorter time for smaller sizes of poultry. Poultry may also be braised on top of the range. Use a heavy, covered pan, add 1/3 cup of water, and follow oven directions. Cook slowly over low to medium heat for 1 1/2 to 2 1/2 hours, adding water if necessary.

Simmering or Stewing

Mature chickens or turkeys need to be simmered or stewed to become tender. Simmer or stew poultry in enough water to barely cover. Add vegetables such as onion and celery if desired. Cover pot and cook over low heat until tender when pierced with a fork. Mature chickens require 1 1/2 to 2 hours; young chickens, 3/4 to 1 hour; and young large turkeys, 3 to 3 1/2 hours.

To shorten cooking time, you can cook in a pressure cooker or pressure saucepan, following the manufacturer's directions.

EGGS

Eggs discussed in this section refer to chicken eggs unless otherwise noted.

Eggs are a nutrient-dense food with a low calorie count. A large egg contains 213 milligrams of cholesterol, but only about 1.5 grams of saturated fat. Egg protein is of exceptionally high quality. The yolk contains almost half of the egg's protein, and, except for magnesium, riboflavin, and niacin, the yolk is higher in nutrients than the egg white. There is no difference in nutrition, taste or functional properties between brown and white-shelled eggs.

SHELL EGGS

Chicken eggs are marketed according to grade and size standards established by USDA or by equivalent standards set by state departments of agriculture. The USDA shield on the egg carton means that the eggs have been federally inspected. Most state laws require that the grade, size, date of pack and/or expiration date appear on the carton label. Plants not under USDA inspection are governed by laws of their states.

Size and grade are two entirely different factors bearing no relationship to one another. The interior and exterior quality of the egg determine grade at the time the egg is packed.

Grades are AA, A and B. Most Grade B eggs go to egg breakers for use in egg products. As the egg ages, the white becomes thinner, allowing the egg to spread more when broken out. This characteristic makes the freshest eggs most desirable when appearance is important, as for poaching or frying. Eggs a week or two old are best for hard cooking, because they can be peeled more easily.

Size is classified according to minimum weight per dozen.

Although eggs are sold by number of eggs per carton, it is easier to determine which size is the best to buy by comparing the price per pound, as shown in Table 6.5.

Weight per Dozen

Jumbo	30 oz.	(850 g)
Extra large	27 oz.	(765 g)
Large	24 oz.	(680 g)
Medium	21 oz.	(595 g)
Small	18 oz.	(510 g)
Peewee	15 oz.	(425 g)

SIZE EQUIVALENTS

Any size egg may be used for frying, scrambling, cooking in the shell or poaching, but other recipes are usually based on the use of large eggs. To substitute another size, use the chart on page 98.

TABLE 6.5

Comparative Prices per Dozen Eggs

Small (18 oz.) (510 g)	Medium (21 oz.) (595 g)	Large (24 oz.) (680 g)	X-Large (27 oz.) (765 g)	Jumbo (30 oz.) (850 g)	Price/lb. (16 oz.) (454 g)
$0.60	$0.70	$0.80	$0.89	$0.99	$0.53
0.64	0.74	0.85	0.95	1.06	0.56½
0.68	0.79	0.90	1.01	1.13	0.60
0.71	0.83	0.95	1.06	1.18	0.63
0.75	0.87	1.00	1.12	1.25	0.66½
0.79	0.92	1.05	1.18	1.31	0.70
0.82	0.96	1.10	1.23	1.37	0.73
0.86	1.00	1.15	1.29	1.43	0.76½
0.90	1.05	1.20	1.35	1.50	0.80
0.93	1.09	1.25	1.40	1.56	0.83
0.97	1.14	1.30	1.46	1.62	0.86½
1.01	1.18	1.35	1.52	1.69	0.90
1.05	1.22	1.40	1.57	1.74	0.93
1.09	1.27	1.45	1.63	1.81	0.96½
1.13	1.31	1.50	1.69	1.88	1.00

TO MAKE 1 CUP

Egg Size	Whole	Whites	Yolks
Jumbo	4	5	11
X-Large	4	6	12
Large	5	7	14
Medium	5	8	16
Small	6	9	18

CARE AND HANDLING

Although a small percentage of shell eggs may contain salmonellae, eggs are a safe food when handled like other perishable foods. Refrigeration is of prime importance in preserving egg quality and safety.

- Buy eggs at retail from refrigerated cases only. For food service, use only eggs delivered under refrigeration.

- Refrigerate with large end up in cartons or cases, at 40°F (4°C) or colder. Properly refrigerated eggs keep 4 to 5 weeks past the pack date without significant loss of quality.

- Use only clean, uncracked eggs.

- Wash cutting boards and all utensils used in egg preparation before using for other foods.

- Do not leave broken-out eggs or egg dishes at room temperature more than two hours including preparation and service. If not served immediately, keep egg dishes below 40°F (4°C) or above 140°F (60°C).

- To store unbroken yolks, cover with water in a tightly covered container, refrigerate, and use in one or two days, or hard cook them, drain well and refrigerate in a tightly sealed container up to four or five days.

- Refrigerate raw whites in a tightly covered container for up to four days.

- Refrigerate hard-cooked unpeeled eggs and use within a week.

FREEZING EGGS

Broken-out eggs can be frozen successfully. Label containers with date, number of eggs, and whether you have added salt or sugar.

Whites—pour into freezer containers and seal tightly.

Yolks—the gelation property of frozen yolks makes them almost impossible to use in a recipe unless 1/8 teaspoon salt (0.5 ml) or 1 1/2 teaspoon (7 ml) sugar or corn syrup per 1/4 cup (60 ml) or 4 yolks is added before freezing.

Whole eggs—beat just until blended and pour into freezer containers. Salt, sugar, or corn syrup needs to be added because of the presence of the yolks.

TO USE FROZEN EGGS

Thaw under running cold water or overnight in the refrigerator. Use immediately. Whites will beat to a greater volume if allowed to stand at room temperature about 30 minutes.

2 Tbsp. (30 ml) thawed egg white = 1 large fresh egg white

1 Tbsp. (15 ml) thawed egg yolk = 1 large fresh yolk

3 Tbsp. (45 ml) thawed whole egg = 1 large fresh egg

Use thawed frozen eggs only in dishes that are thoroughly cooked.

MICROWAVE COOKING

Because few foods are as heat-sensitive as eggs, a reduced power level of 200 to 300 watts is recommended for all egg dishes except omelets, poached eggs with water added, and scrambled eggs.

Do not cook raw eggs in their shells or reheat peeled or unpeeled hard-cooked eggs. They will explode. Rapid build-up of steam can cause yolks of broken-out eggs to explode. Always prick yolks of unbeaten eggs and cover container with waxed paper or plastic wrap to avoid splattering.

Microwave cooking is not recommended for soufflés and puffy omelets.

For quick hard-cooked eggs, separate yolks and whites into two small bowls. Stir yolks. Cover and cook separately on reduced power (200 to 300 watts), stirring once or twice. Allow 20 to 30 seconds per yolk and about 30 seconds to 1 minute per white. Let stand, covered, about 2 minutes. Cool long enough to handle, then chop or chill until ready to use.

SAFETY OF EGG-CONTAINING DISHES

Salmonellae will not survive in dishes cooked to 140°F (60°C) for at least 3½ minutes or in those cooked to 160°F (80°C). Cook poached, fried, soft-cooked, and scrambled eggs until whites are completely coagulated and yolks begin to thicken. Cook omelets, frittatas, and French toast until no visible liquid remains. When serving infants, the elderly, or the ill or immune-compromised, cook all egg dishes thoroughly or use pasteurized egg product.

WHIPPING PROPERTIES

Beaten egg whites provide the volume and texture of many baked products, puffy omelets, soufflés and chiffon cakes. The volume of beaten egg whites can increase 6 to 8 times.

Fat inhibits foam formation, therefore separate yolks and whites carefully to avoid the presence of any yolk in the white. Use metal or glass bowls for beating whites. Plastic bowls cannot be rendered completely fat free.

Separate eggs and beat egg whites cold. Egg white foams should be beaten just until whites no longer slip when the bowl is tilted. Under beaten whites result in poor volume of the finished product. Over beaten whites have lost their elasticity and break apart when combined with other ingredients.

Drainage occurs fairly rapidly so whites should not be beaten more than a few minutes before they are needed.

Adding ⅛ teaspoon (0.5 ml) cream of tartar per egg white helps to stabilize the foam.

Sugar should be beaten in 1 or 2 teaspoons (5 or 10 ml) at a time after foaming has begun.

Beaten whole egg or egg yolk will increase about four times in volume. Beat until eggs become lemon-colored and form ribbons when beater is lifted, about three to five minutes. Egg yolks cannot be overbeaten as whites can be.

EGG PRODUCTS

Processed and convenience forms of eggs available to food service include refrigerated liquid, frozen, dried (egg solids), and specialty products. All egg products are produced in federally inspected plants and are pasteurized so they are free of salmonella.

TABLE 6.6

Large Shell Egg Equivalency to Frozen or Refrigerated Liquid Egg Products

		Weight		Measure	
Whole eggs					
	1	1¾ oz.	50 g	3 Tbsp.	45 ml
	10	1 lb. 1¾ oz.	503 g	2 cups	473 ml
	12	1 lb. 5½ oz.	610 g	2½ cups	592 ml
	25	2 lbs. 13 oz.	1276 g	1 qt. 1¼ cups	1242 ml
	50	5 lbs. 8 oz.	2495 g	2 qt. 2½ cups	2483 ml
Yolks					
	10	7¼ oz.	206 g	¾ cup	177 ml
	12	8½ oz.	241 g	¾ cup 2 Tbsp.	207 ml
	22	1 lb.	454 g	2 cups less 2 Tbsp.	443 ml
Whites					
	10	11½ oz.	326 g	1¼ cups 2 Tbsp.	326 ml
	12	14 oz.	397 g	1½ cups 2 Tbsp.	385 ml
	14	1 lb.	454 g	2 cups less 1 Tbsp.	458 ml

TABLE 6.7

Shell Egg Equivalency to Dried Whole Egg Products

Shell Egg (Large Size)		Dried Whole Sifted		Water	
6	3 oz.	85 g	1 cup	237 ml	1 cup
12	6 oz.	170 g	2 cups	473 ml	2 cups
24	12 oz.	340 g	1 qt.	946 ml	1 qt.
50	1 lb. 9 oz.	709 g	2 qt. 1/3 cup	1971 ml	2 qt. 1/3 cup
100	3 lb. 2 oz.	1418 g	1 gal. 2/3 cup	3942 ml	1 gal. 2/3 cup
150	4 lb. 11 oz.	2126 g	6 qt. 1 cup	5915 ml	6 qt. 1 cup
200	6 lb. 4 oz.	2835 g	2 gal. 11/3 cup	7886 ml	2 gal. 11/3 cup

Retail markets carry egg substitutes, in which yolks are replaced with oil, dried milk, gums or other ingredients. Dried whole eggs usually are available at camping goods stores, and meringue powder, based on dried egg white, is available at craft stores since it is frequently used to frost baked goods. Equivalency of large shell eggs to frozen, refrigerated liquid, and dried egg products is shown in Table 6.6 and Table 6.7.

HANDLING EGG PRODUCTS

Do not accept frozen egg products that are beginning to thaw. Transfer frozen products to the freezer immediately after delivery.

Do not thaw at room temperature. Thaw in the refrigerator or under cold running water in sealed containers. Use defrosted egg products promptly. Cover and refrigerate unused portion and use within three days.

Refrigerate liquid egg products immediately upon delivery. Liquid product will keep, refrigerated at 40°F (4°C) or below and unopened, up to six days. Once opened, use immediately.

Store dried egg products in a dark cool (below 70°F, 20°C) place, preferably in the refrigerator.

Reconstitute only the amount of dried eggs that will be used immediately. Sift before measuring.

After opening dried eggs, seal tightly for re-storage and keep below 50°F (10°C).

Keep specialty egg products refrigerated or frozen according to their requirements.

Fruits

FRESH FRUITS

Many varieties of fresh fruits are available all year round in almost every part of the United States because of excellent transportation and storage facilities.

KINDS OF FRUITS

Fruits may be classified as follows:

Berries—Those commonly marketed are blackberries, blueberries, boysenberries, cranberries, currants, elderberries, gooseberries, grapes, raspberries and strawberries.

Citrus—These are the grapefruits, kumquats, lemons, limes, mandarins, oranges, tangerines and tangelos.

Drupes—This is the botanical name given to apricots, cherries, nectarines, peaches, plums and prunes; also known as stonefruits.

Melons—These vine-grown fruits include cantaloupe, casaba, Crenshaw, honeydew, Persian and watermelon.

Pomes—Fruits include apples, pears, prickly pears and quinces.

Tropical and Semi-Tropical Fruits—There are many species of tropical and semi-tropical fruits but those that are marketed commonly throughout the United States are avocados, bananas, coconuts, dates, figs, guavas, mangoes, papayas, persimmons, pineapples and pomegranates. More exotic fruits are Asian pear, atemoya, breadfruit, canistel, carambola (star fruit), cherimoya (custard apple), feijoa, jaboticaba, longans, lychees, mamey, monstera, passion fruit (granadilla), quince, sapodilla, sapote, tamarinds and ugli fruit.

GRADES

Grades of fresh fruits are established by the USDA's Agricultural Marketing Service (AMS) and used primarily in wholesale channels of distribution. They have been widely used by growers, shippers and car lot receivers for domestic and foreign shipment but have not been so extensively used in the retail trade as those for canned products which are packed in consumer-size containers. However, consumers have made considerable use of U.S. grades in the purchase of such commodities such as apples, grapefruit, oranges and peaches. These voluntary grades frequently are found on boxes of fresh fruit.

In the U.S., the principal factors used as standards for grades of fresh fruit are maturity; decay, if any; shipping quality; appearance; and waste caused by various defects. The U.S. No. 1 grade is designed to include a fairly good proportion of the commercial crop. Standards for some commodities provide a Fancy or Extra No. 1 grade for use of packers of superior fruits for which a premium is obtained. Apples, for example, may be graded U.S. Extra Fancy, U.S. Fancy, U.S. No. 1, or combination of these grades. There are also lower U.S. grades for fresh fruits, such as No. 2, No. 1, Cookers, and Combination. Tropical fruits currently are not marketed under U.S. grades.

SIZES

Fruits sold in quantity are usually sold by weight or count rather than size. Medium sizes tend to be more desirable than very large, which often lack quality, or very small, which have little edible portion compared to wasted parts. At retail, many fruits that used to be sold by number may now be sold by weight. With the development in biotechnology, some fruits have been downsized to create miniatures, such as melons.

A FEW REMINDERS

- Apple variations for eating out of hand or in salads or fruit cups include Delicious, Gala, Golden Delicious, Jonathan, Macintosh, Spartans, Stayman and Winesap. The tart and slightly acid apples—Granny Smith, Gravenstein, Grimes Golden, Jonathan, and Newton— are excellent for pies and sauces. Firm-fleshed apples that hold shape well in baking include Northern Spy, Rhode Island Greening, Rome Beauty, Winesap, and York Imperial.

- Cherries may be of the sweet variety for eating out of hand or they may be tart, sometimes called sour cherries, and used in cooking and baking. These are lighter red in color than the sweet cherries and have a softer flesh. Most of the commercial crop of tart cherries goes to processors.

- Figs may be black, yellow or green. For eating fresh, the black figs are popular.

- The European varieties of grapes include Cardinal (early bright red), Emperor (late deep red), and Ribier (late black), Thompson seedless (early green) and Tokay (late red). The American varieties include Catawba, Concord (blue-black), Delaware and Niagara. Varietal groups include table, raisin, wine, juice and canning.

- Grapefruit may be white- or pink-fleshed, with or without seeds.

- Limes may be available in two basic varieties. The Persian or Tahiti lime is rounder, larger, and thinner skinned than the Key lime (also known as the Mexican or West Indies lime). As limes ripen, they become yellow and less acid.

- Of the melon family, the mature cantaloupe has no stem but a slight indention at the stem end and a yellow to light green netting. Casaba melon is pumpkin shaped and has no netting but does have shallow, irregular, lengthwise furrows. Honeydew melon is large, bluntly oval, and smooth. Persian melon resembles cantaloupe but is nearly round, has finer netting and is similar in size to honeydew. Santa Claus or Christmas melon looks like a small watermelon with the flesh of a honeydew. Crenshaw melon is oval-shaped with a deep golden rind and no netting.

- Orange varieties include Navel from California and Arizona; Valencia from California, Arizona, Florida and Texas; and mandarins. Mandarins include satsumas, tangerines and miscellaneous hybrids, such as Temple oranges from Florida, a cross between a tangerine and an orange. Murcott is a cross between a sweet orange and a mandarin. Tangelo is a hybrid of a tangerine or mandarin orange and a grapefruit or pomelo.

- Peach varieties are divided into freestone (Elberta, O'Henry and Redtop) and clingstone (Flavorcrest, June Lady, Red Haven and Springcrest). As the names imply, the flesh of the freestone peach separates readily from the pit, whereas the reverse is true of the clingstone. In general, freestones are preferred for eating fresh and for freezing. Both kinds may be canned, but the clingstone maintains the firmer shape.

- Bartlett pears, Forelle and Seckel are summer and early fall pears that are eaten fresh and are also used in canning. Fall and winter pear varieties include Bosc, D'Anjou, Comice and Winter Nellis. These keep well in cold storage.

- Pineapples are picked at the peak of ripeness and flavor; they do not ripen after harvest.

- Plums may be red, green or purplish yellow. Prunes are a variety of freestone plums suitable for drying; they are purplish-black and smaller and more oval than other types of plums.

STORAGE AND USE

- Store citrus fruits—except tangerines—in the refrigerator uncovered. Place tangerines in a plastic bag for refrigerator storage.

- Refrigerate ripe stonefruits uncovered and plan to use within three to five days.

- Store ripe melons in a plastic bag to protect other foods from the pungent melon odor and store in refrigerator for use within a week.

- Keep bananas at room temperature until ripe and then refrigerate. Cold temperatures darken the skins, but will not affect palatability.

- Let firm avocados stand three to five days until they soften; then refrigerate.

- Coat cut fruit with lemon juice or ascorbic acid to avoid discoloration of pared apples, peaches, bananas and avocado that occurs when the pared fruits are exposed to air.

- To ripen cantaloupe, let stand at room temperature two to four days. Chill a few hours in refrigerator before serving.

- Store late-maturing apple varieties, like Jonathan and Red Delicious, for 3–5 months if harvested at peak ripeness without bruising. Refrigerate immediately after picking at 30–32°F with 90% humidity. Perforated plastic bags will allow for some air circulation.

- Pick pears when still slightly green and firm. Refrigerate at 30–35°F for 1 to 3 months. To ripen, place at room temperature for 1 week. To ripen faster, seal pears in plastic bag; accumulated ethylene gas will hasten the process.

- Most tropical fruits will soften at room temperature, giving slightly to pressure when ripe. Ripe fruit needs to be refrigerated.

PROCESSED FRUITS

Many fruits are available in frozen, canned, dried and dehydrated forms. The number and kinds of processed foods available at the retail market are constantly increasing as new procedures and equipment are developed for processing foods of improved quality. Their convenience and availability all year round have added variety to menus particularly in winter.

FROZEN AND CANNED FRUITS

Standards of identity for most canned fruits as well as minimum standards of quality for many of the principal ones have been established by the U.S. Food and Drug Administration. The U.S. Department of Agriculture has developed grade standards that also reflect factors of quality for many frozen and canned fruits. Canners, freezers and distributors often voluntarily use the grade designation on their labels. Grades reflect mainly differences in appearance. The factors most important in evaluating the quality of processed fruits are flavor, color, texture, uniformity of size and shape, ripeness, and absence of defects. The permissive grades established by industry for most processed fruits are A (fancy), B (choice) or C (standard). The terms in parentheses are those in common commercial usage for designating the quality of processed products.

DRIED FRUITS

Apples, apricots, blueberries, cherries, figs, peaches, prunes and raisins may be sun dried, mechanically dehydrated, vacuum dried, foam-mat dried or explosively puffed. The moisture content of the dried fruits generally ranges from 15 to 25 percent. Quality levels or grades

for most dried fruit are the same as for other fruits. The sizes of dried fruit—small, medium, large, extra large or jumbo—appear on retail packages.

Prunes are size graded according to the number per pound, with size grading varying somewhat with type— French, Italian, Imperials and/or Sugars.

When "sulfur dioxide" appears on the label of dried apples, apricots and peaches, the fruit may have been dipped in sodium bisulfate or treated with sulfur dioxide fumes to prevent darkening of color. The chemical is harmless and disappears in the steam when the fruits are cooked.

Fruit leathers are prepared by dehydrating a puree of selected fruit and sweetener. All dried fruits, including leathers, must be stored in airtight packaging to retain quality and avoid the addition of moisture.

FRUIT JUICES

Juice may be extracted from virtually any fruit and may be purchased as full strength or concentrate to be reconstituted. In addition to fresh and frozen fruit juices, canned, frozen and powdered fruit drinks are available. Fruit drinks are prepared from regular or synthetic fruit juices or from a combination of the two and contain natural or synthetic flavorings, sugar or non-sugar sweetener, color, usually ascorbic acid and preservatives. Nutritionally, fruit juices are a better choice than fruit drinks. The National Juice Products Association has adopted a voluntary policy to label dilute juice beverages for juice content as "contains ___ percent of ___ juice."

QUALITY GUIDE FOR FRUITS

APPLES

What To Look For: Firm, well-colored fruit. Apples must be mature when picked to have good flavor, crisp texture and storing ability.

What To Avoid: Immature apples that lack color for the particular variety; also fruit with shriveled skin; overripe apples (indicated by a yielding to slight pressure on the skin and soft mealy flesh); and apples affected by freezing (indicated by internal breakdown and bruised areas). Scald on apples (irregularly shaped tan or brown areas) may not seriously affect the eating quality of the apple.

APRICOTS

What To Look For: Apricots that are plump and juicy looking and have a uniform, golden-orange color. Ripe apricots will yield to gentle pressure on the skin.

What To Avoid: Dull-looking, soft or mushy fruit, and very firm, pale yellow or greenish yellow fruit. These are indications of over-maturity or immaturity, respectively.

AVOCADOS

What To Look For: For immediate use, slightly soft avocados that yield to a gentle pressure on the skin. For use in a few days, firm fruits that do not yield to the squeeze test. Leave them at room temperature to ripen. Irregular light brown markings are sometimes found on the outside skin. These markings have no effect on the flesh of the avocado.

What To Avoid: Avocados with dark sunken spots in irregular patches or cracked or broken surfaces. These are signs of decay.

BANANAS

What To Look For: Bananas that are firm, bright in appearance and free from bruises or other injury. The stage of ripeness is indicated by the skin color, best eating quality has been reached when the solid yellow color is specked with brown. At this stage, the flesh is mellow and the flavor is fully developed. Bananas with green tips or with practically no yellow color have not developed their full flavor potential.

What To Avoid: Bruised fruit (which means rapid deterioration and waste); discolored skins (a sign of decay); a dull, grayish, aged appearance (showing that the bananas have been exposed to cold and will not ripen properly). Occasionally, the skin may be entirely brown and yet the flesh will still be in prime condition. Flavor will be fully developed.

BLUEBERRIES

What To Look For: A dark blue color with a silvery bloom which is a natural, protective waxy coating; blueberries that are plump, firm, uniform, dry and free from stems or leaves.

What To Avoid: Soft, mushy berries; berries with broken skins; moldy berries.

CANTALOUPES (MUSKMELONS) AND PERSIAN MELONS

What To Look For: Three major signs of full maturity:

1. The stem should be gone, leaving a smooth, symmetrical, shallow basin called a "full-slip." If all or part of the stem base remains or if the stem scar is jagged or torn, the melon is probably not fully matured.

2. The netting, or veining, should be thick, coarse and corky and should stand out in bold relief over some part of the surface.

3. The skin color (ground color) between the netting should have changed from green to a yellowish buff, yellowish gray or pale yellow.

Look for signs of ripeness, for a cantaloupe may be mature, but not ripe. A ripe cantaloupe will have a yellowish cast to the rind, have a pleasant cantaloupe odor when held to the nose, and will yield slightly to light thumb pressure on the blossom end of the melon.

What To Avoid: Over-ripeness, shown by a pronounced yellow rind color, a softening over the entire rind, and soft, watery and insipid flesh. Small bruises normally will not hurt the fruit, but large bruised areas should be avoided since they generally cause soft, water-soaked areas underneath the rind. Mold growth on the cantaloupe—particularly in the stem scar with wet tissue under the mold—indicates decay.

CASABA

What To Look For: Ripe melons with a gold yellow rind color and a slight softening at the blossom end. Casabas have no odor or aroma.

What To Avoid: Decayed melons, shown by dark, sunken water-soaked spots.

CHERRIES

What To Look For: A very dark color, the most important indication of good flavor and maturity in sweet cherries. Bing, Black Tartarian, Chapman, Republican and Schmidt varieties should range from deep maroon to mahogany red to black, for richer flavor. Lambert cherries should be dark red. Good cherries have bright, glossy, plump-looking surfaces and fresh looking stems.

What To Avoid: Over-mature cherries lacking in flavor, indicated by shriveling, dried stems, and a generally dull appearance. Decay is fairly common at times on sweet cherries, but because of the normal dark color, decayed areas are often inconspicuous. Soft, leaking fruit, brown discoloration, and mold growth are indications of decay.

CRANBERRIES

What To Look For: Plump, firm berries with a lustrous color for the best quality. Duller varieties should at least have some red color. Occasional soft, spongy, or leaky berries should be sorted out before cooking because they may produce an off-flavor.

CRENSHAW

What To Look For: A deep golden yellow rind, sometimes with small areas of a lighter shade of yellow; a surface that yields slightly to moderate pressure of the thumb, particularly at the blossom end; a pleasant aroma.

What To Avoid: Slightly sunken, water-soaked areas on the rind (a sign of decay, which spreads quickly through the melon).

GRAPEFRUIT

What To Look For: Firm, well-shaped fruits—heavy for their size—are usually the best eating. Thin-skinned fruits have more juice than coarse-skinned ones. If a grapefruit is pointed at the stem end, it is likely to be thick-skinned. Rough, rigid or wrinkled skin can also be an indication of thick skin and lack of juice.

Grapefruit often has skin defects—such as scale, scars, thorn scratches, or discoloration—which usually do not affect the eating quality.

What To Avoid: Soft, discolored areas on the peel at the stem end; water-soaked areas; loss of bright color; and soft and tender peel that breaks easily with finger pressure. These are all symptoms of decay, which has an objectionable effect on flavor.

GRAPES

What To Look For: Well-colored plump grapes that are firmly attached to the stem. White or green grapes are sweetest when the color has a yellowish cast or straw color, with a tinge of amber. Red varieties are better when a good red predominates on all or most of the berries. Bunches are more likely to hold together if the stems are green and pliable. Grapes will not improve in sweetness, quality or color after picking.

What To Avoid: Soft or wrinkled grapes (showing effects of freezing or drying), grapes with bleached areas around the stem end (indicating injury and poor quality), and grapes leaking brownish fluid (a sign of decay).

HONEYBALL AND HONEYDEW

What To Look For: Maturity, shown by a soft, velvety feel, and for ripeness, shown by a slight softening at the blossom end, a faint pleasant fruit aroma, and a yellowish white to cream rind color.

What To Avoid: Melons with a dead-white or greenish-white color and hard, smooth feel (which are signs of immaturity), large, water-soaked bruised areas (sign of injury), and cuts or punctures through the rind (which usually lead to decay). Small, superficial, sunken spots do not damage the melon for immediate use but large decayed spots will.

KIWIFRUIT

What To Look For: Kiwifruit is ripe when it yields to gentle pressure. To ripen, place hard fruit in a plastic bag with an apple or banana for one to two days at room temperature.

What To Avoid: Soft dark areas, indentations, broken skin, uneven color.

LEMONS

What To Look For: Lemons with a rich yellow color, reasonably smooth-textured skin with a slight gloss, and those that are firm and heavy. A pale or greenish yellow color means very fresh fruit with slightly higher acidity. Coarse or rough skin texture is a sign of thick skin and not much flesh.

What To Avoid: Lemons with a darker yellow or dull color, or with hardening or shriveling of the skin (sign of age) and those with soft spots, mold on the surface, and punctures of the skin (sign of decay).

LIMES

What To Look For: Limes with a glossy skin and heavy weight for the size.

What To Avoid: Limes with full, dry skin (a sign of aging and loss of acid flavor), and those showing evidence of decay (soft spots, mold and skin puncture).

NECTARINES

What To Look For: Rich color and plumpness and a slight softening along the "seam" of the nectarine. Most varieties have an orange-yellow color (ground color) between the red areas, but some varieties have a greenish ground color. Bright-looking fruits which are firm to moderately hard will probably ripen normally within two to three days at room temperature.

What To Avoid: Hard, dull fruits or slightly shriveled fruits (which may be immature—picked too soon—and of poor eating quality), and soft to overripe fruits or those with cracked or punctured skin or other signs of decay. Russeting or staining of the skin may affect the appearance but not detract from the internal quality of the nectarine.

ORANGES

What To Look For: Firm and heavy oranges with fresh, bright-looking skin that is reasonably smooth for the variety.

What To Avoid: Lightweight oranges, which are likely to lack flesh content and juice. Very rough skin texture indicates abnormally thick skin and less flesh. Dull, dry skin and spongy texture indicate aging and deteriorated eating quality. Also avoid decay—shown by cuts or skin punctures, soft spots on the surface, and discolored, weakened areas of skin around the stem end or button.

PEACHES

What To Look For: Peaches which are fairly firm or becoming a trifle soft. The skin color between the red areas (ground color) should be yellow or at least creamy with a peachy fragrance.

What To Avoid: Very firm or hard peaches with a distinctly green ground color, which are probably immature and will not ripen properly. Also avoid very soft fruits, which are overripe. Do not buy peaches with large flattened bruises (they will have areas of discolored flesh underneath) or peaches with any sign of decay. Decay starts as a pale tan spot that expands in a circle and gradually turns darker in color.

PEARS

What To Look For: Firm pears of all varieties. The color depends on variety. For Bartletts, look for a pale yellow to rich yellow color; Bosc—greenish yellow to brownish yellow (the brown cast is caused by skin russeting, a characteristic of the Bosc pear); D'Anjou or Comice—light green to yellowish green; Winter Nellis—medium to light green.

Pears which are hard when purchased will probably ripen if kept at room temperature, but it is wise to select pears that have already begun to soften—to be reasonably sure that they will ripen satisfactorily.

What To Avoid: Wilted or shriveled pears with dull-appearing skin and slight weakening of the flesh near the stem, which indicates immaturity. These pears will not ripen. Also avoid spots on the sides or blossom ends of the pear, which means that corky tissue may be underneath.

PINEAPPLES

What To Look For: The color, the fragrant pineapple odor, a very slight separation of the eyes or pips, and the ease with which the "spikes" or leaves can be pulled out from the top. Pineapples are usually dark green in mature hard stage. As the more popular varieties (such as

Red Spanish and Smooth Cayenne) ripen, the green color fades and orange and yellow take its place. When fully ripe, the pineapples are golden yellow, orange yellow or reddish brown—depending on the variety, although one seldom-seen pineapple (the Sugar Loaf) remains green even when ripe. Also look for maturity, shown by plump glossy eyes or pips, firmness, a lively color, and fruits which are heavy for their size.

What To Avoid: Pineapples with sunken or slightly pointed pips, dull yellowish-green color, and dried appearance (all signs of immaturity). Also avoid bruised fruit (shown by discolored or soft spots), which are susceptible to decay. Other sign of decay (which spreads rapidly through the fruit) are traces of mild, unpleasant odor, and eyes that turn watery and darken in color.

PLUMS AND PRUNES

What To Look For: Plums and prunes with a good color for the variety, in a fairly firm to slightly soft stage of ripeness.

What To Avoid: Fruits with skin breaks, punctures or brownish discoloration. Also avoid immature fruits (relatively hard, poorly colored, very tart, sometimes shriveled) and over-mature fruits (excessively soft, possibly leaking or decaying).

RASPBERRIES, BOYSENBERRIES, GOOSEBERRIES, LOGANBERRIES

What To Look For: A bright, clean appearance and a uniform good color for the species. The individual small cells making up the berry should be plump and tender but not mushy. Look for berries that are fully ripened and with no attached stem caps.

What To Avoid: Leaky and moldy berries. Also note wet or stained spots on wood or fiber containers; these are possible signs of poor quality or spoiled berries.

STRAWBERRIES

What To Look For: Berries with a full red color and a bright luster, firm flesh, and the cap stem still attached. The berries should be dry and clean; usually medium to small strawberries have better eating qualities than do the larger berries.

What To Avoid: Berries with large uncolored areas or with large seedy areas (poor in flavor and texture), a dull shrunken appearance or softness (sign of decay), or those with mold, which can spread rapidly from one berry to another.

TANGERINES

What To Look For: Deep yellow or orange color and a bright luster are the best sign of fresh, mature, good-flavored tangerines. Because of the typically loose nature of the tangerine skin, they will frequently not feel firm to the touch.

What To Avoid: Very pale yellow or greenish fruits, likely to be lacking in flavor (although small green areas on otherwise high-colored fruits are not bad) and tangerines with cut or punctured skins or very soft spots (all signs of decay, which spreads rapidly).

WATERMELON

What To Look For: In cut melons, firm, juicy flesh with good red color, free of white streaks; seeds which are dark brown or black. In whole melons, a relatively smooth surface, neither shiny nor dull, ends that are rounded and filled out; creamy colored underside.

What To Avoid: Melons with pale-colored flesh, and white streaks or "white heart," whitish seeds (indicating immaturity). Dry, mealy flesh or watery, stringy flesh are signs of over-maturity or aging after harvest.

Vegetables

FRESH VEGETABLES

Fresh vegetables are available throughout the year due to modern transportation and storage. U.S. markets now sell fresh vegetables grown around the world. Vegetables are classified according to the portion of the plant that is eaten. The following classification is a general list; many more types and varieties are available worldwide.

Bulbs: Onions (all varieties), garlic, leeks and shallots.

Flowers and Fruits: Artichoke, broccoli, cauliflower, chayote, cucumber, eggplant, fennel, nopales, okra, pepper, pumpkin, squash, sweet corn, tomato and water chestnuts.

Leaves and Stems: Asparagus, Brussel sprouts, cabbage, celery, Chinese cabbage, chives, greens, kale, kohlrabi, lettuce, parsley, seaweed, spinach and watercress.

Legumes: USDA describes legumes as the mature, dry seeds of the family Fabaceae or Leguminosae. The list includes: Adzuki, common beans, broadbean (Vicia faba), carob, chickpeas, cowpeas, hyacinth beans (dahl), lentils, lima beans, lupins, mothbeans, mung beans, mungo beans, peas, peanuts, pigeon peas, soybeans, and winged beans.

Roots: Beet, carrot, cassava (manioc or tapioca), celeriac, daikon, jicama, kohlrabi, poi (taro), parsnip, radish, rutabaga, salsify, sweet potato and turnip.

Tubers: Jerusalem artichoke, Irish potato and taro.

Bamboo shoots, mushrooms and truffles are also considered to be vegetables.

CRUCIFERS

Crucifers may provide even more health benefits than other vegetables. Crucifers include Bok choy, broccoli, Brussel sprouts, cabbage, cauliflower, collards, kale, kohlrabi, mustard greens, rutabagas, turnips and their greens.

GRADES FOR FRESH VEGETABLES

U.S. grading standards for fresh vegetables are voluntary; the principal factors affecting the grade are maturity, decay, shipping quality, appearance and waste caused by various defects. The U.S. No. 1 grade is designed to include a fairly good proportion of the commercial crop. Standards for some commodities provide a fancy grade used by packers of superior vegetables for which a premium is obtained. U.S. No. 1 is the highest grade for most vegetables. Other fresh vegetable grades (No. 2 or Combination) are not likely to appear in retail stores.

CANNED AND FROZEN VEGETABLES

Vegetables may be frozen or canned separately or in combination with other vegetables. The different forms include whole, sliced, diced, mashed, puree, juice, and soup.

GRADES FOR CANNED AND FROZEN VEGETABLES

The Standards of Quality and Fill as outlined by FDA are minimal, basic requirements that serve as a pattern for good processing techniques for the commercial food processor. FDA Standards of Identity tell us what a product is.

The U.S. Grade Standards developed by USDA are essentially market classifications of quality. Processors are not required by law to include the grade on the label. The U.S. grade names are U.S. Grade A or Fancy, U.S. Grade B or Extra-Standard, and U.S. Grade C or Standard. The four quality characteristics used to determine the grade are color, uniformity of size, texture, and absence of defects. Additionally grading factors that are used include consistency, clearness of liquid, size and symmetry, drained weight, wholeness and flavor.

STORAGE OF CANNED AND FROZEN VEGETABLES

Store unopened canned vegetables in a cool, dry place. For best quality, use canned vegetables within a year of purchase. Vegetables lose quality and some nutrients if stored too long, but remain indefinitely safe to eat if the seal is not broken and the can is not bulging.

Frozen vegetables should be stored at 0°F (-18°C) or lower to retain quality and nutritive value for 8 to 12 months. Before buying large qualities of frozen vegetables check the freezer or freezing compartment temperature. If the temperature is above 0°F (-18°C), plan to use frozen vegetables within a few days for best quality.

DRIED AND FREEZE-DRIED VEGETABLES

Dried and freeze-dried vegetables are readily available in markets, but health food and sporting goods sections in large stores may carry a wide selection of dried and freeze-dried foods.

STORAGE OF DRIED VEGETABLES

Optimal storage conditions for dried foods are:

- Storage temperature below 60°F
- Cool, dry, and dark storage area, and
- Airtight and moisture-proof packaging.

For best quality and nutritional value, dried foods should be used within one year of purchase.

SUBSTITUTING DRIED FOODS FOR FRESH

Substitute reconstituted vegetable powders and flakes for finely chopped fresh vegetables in recipes. If you do not reconstitute the powders or flakes before adding them to the recipe ingredients, add a small amount of liquid to the recipe. Allow 1/4 cup (60 ml) liquid for each tablespoon of powder and each 1 1/2 tablespoons (23 ml) of flakes.

Substitute 1 tablespoon (15 ml) vegetable powder or 1 1/2 tablespoons vegetable flakes or 2 tablespoons (30 ml) dried pieces for 4 tablespoons (60 ml) chopped fresh vegetables.

QUALITY GUIDE FOR VEGETABLES

ARTICHOKES

What To Look For: Plump, globular artichokes that are heavy in relation to size, and compact with thick, green, fresh looking leaf scales. Size is not important in relation to quality.

What To Avoid: Artichokes with large areas of brown on the scales and with spreading scales (a sign of age, indicating drying and toughening of the edible portions), grayish black discoloration (caused by bruises), mold growth on the scales, and worm injury.

ASPARAGUS

What To Look For: Closed, compact tips, smooth, round spears, and a fresh appearance. A rich green color should cover most of the spear. Stalks should be tender almost as far down as the green extends.

What To Avoid: Tips that are open and spread out, moldy or decayed tips, or ribbed spears (up and down ridges), or spears that are not rounded. These are signs of aging and indicate tough asparagus and poor flavor. Also avoid excessively sandy asparagus because sand grains can lodge beneath the scales and are difficult to wash out.

BEAN (SNAP BEANS)

What To Look For: A fresh, bright appearance with good color for the variety. Get young, tender beans with pods in firm crisp condition.

What To Avoid: Wilted or flabby bean pods, serious blemishes, and decay. Thick, tough, fibrous pods indicate over-maturity.

BEETS

What To Look For: Beets that are a rich, deep red color, firm, round and smooth over most of the surface and have a slender tap root (the large main root). If beets are bunched, judge their freshness by the condition of the tops. Badly wilted or decayed tops indicate a lack of freshness, but the roots may be satisfactory if they are firm.

What To Avoid: Elongated beets with round, scaly areas around the top surface; these will be tough, fibrous and strong flavored. Also avoid wilted, flabby beets which have been exposed to the air too long.

BROCCOLI

What To Look For: A firm, compact cluster of small flower buds, with none opened enough to show the bright yellow flower. Bud clusters should be dark green or sage green or even green with a decidedly purplish cast. Stems should not be too thick or tough.

What To Avoid: Broccoli with spread bud clusters, enlarged or open buds, yellowish green color or wilted condition—signs of over-maturity and overlong display. Also avoid broccoli with soft, slippery, water soaked spots on the bud cluster. These are signs of decay.

BRUSSEL SPROUTS

What To Look For: A fresh, bright green color, tight fitting outer leaves, firm body, and freedom from blemishes.

What To Avoid: Brussel sprouts with yellow or yellowish green leaves or leaves that are loose, soft or wilted. Small holes or ragged leaves may indicate worm injury.

CABBAGE

Cabbage may be red, green with smooth leaves, or green with crinkly leaves—a variety known as Savoy.

What To Look For: Firm or hard heads of cabbage that are heavy for their size. Outer leaves should be a good green or red color (depending on type), reasonably fresh, and free from serious blemishes. The outer leaves (called "wrapper" leaves) fit loosely on the head and are usually discarded, but too many loose wrapper leaves on a head cause waste. With Savoy cabbage, look for crumpled leaves with developed round heads of dark green color.

Some early crop cabbage may be soft or only fairly firm but is suitable for immediate use if the leaves are fresh and crisp. Cabbage that has been in storage is usually trimmed of all outer leaves and lacks color but is satisfactory if not wilted.

What To Avoid: New cabbage with wilted or decayed outer leaves or with leaves turning yellow. Worm-eaten outer leaves often indicate that the worm injury penetrates into the head.

Stored cabbage with outer leaves badly discolored, dried or decayed probably is over-aged. Separation of the stems of leaves from the central stem at the base of the head also indicates overage.

CARROTS

What To Look For: Carrots that are well formed, smooth, well colored and firm.

What To Avoid: Roots with large green sunburned areas at the top (which must be trimmed) and roots that are flabby from wilting or show spots of soft decay.

CAULIFLOWER

What To Look For: White to creamy compact solid and clean curds (head). A slightly granular texture of the curd will not hurt the eating quality if the surface is compact. Ignore small green leaflets extending through the curd. If jacket leaves are attached, a green color is a sign of freshness.

What To Avoid: A spreading of the curd—a sign of aging or over-maturity. Also avoid wilting or discolored spots on the curd. A smudgy or speckled appearance of the curd is a sign of insect injury, mold growth or decay and should be avoided.

CELERY

What To Look For: Freshness and crispness: the stalk should have a solid, rigid feel, and leaflets should be fresh or only slightly wilted. Also look for a glossy surface, light or medium green stalks and green leaflets.

What To Avoid: Wilted celery with flabby upper branches or leaf stems. Celery with pithy, hollow, or discolored centers in the branches. Celery with internal discoloration will show some gray or brown on the inside surface of the larger branches near the base of the stock. Avoid celery with "blackheart," a brown or black discoloration of the small center branches; insect injury in the center branches or the insides of outer branches; long, thick seedstem in place of the usually small, tender heart branches.

CHARD (see GREENS)

CHICORY, ENDIVE, ESCAROLE

What To Look For: Freshness, crispness and tenderness: look for a good green color of the outer leaves (except for Witloof or Belgian endive, the compact, cigar-shaped creamy white endive grown in complete darkness).

What To Avoid: Plants with leaves which have brownish or yellowish discoloration or which have insect injury.

CHINESE CABBAGE

What To Look For: Fresh, crisp, green plants that are free from blemishes or decay.

What To Avoid: Wilted or yellowed plants.

COLLARDS (see GREENS)

CORN

What To Look For: Fresh, succulent husks with good green color, silk ends that are free from decay or work injury, and stem ends (opposite from the silk) that are not too discolored or dried. Select ears that are well covered with plump, not too mature kernels.

What To Avoid: Ears with undeveloped kernels which lack yellow color (in yellow corn), ears with very large kernels, and ears with dark yellow kernels with depressed areas on the outer surface. Also avoid ears of corn with yellowed, wilted, or dried husks, or discolored and dried out stem ends.

CUCUMBERS

What To Look For: Cucumbers with good green color which are firm over their entire length. They should be well shaped and well developed, but should not be too large in diameter. Good cucumbers typically have small lumps on their surfaces. They may also have some white or greenish white color and still be of top quality.

What To Avoid: Overgrown cucumbers which are large in diameter and have a dull color, turning yellowish. Also avoid cucumbers with withered or shriveled ends—signs of toughness and bitter flavor.

EGGPLANT

What To Look For: Firm, heavy, smooth and uniformly dark purple eggplants.

What To Avoid: Those which are poorly colored, soft, shriveled, cut or which show decay in the form of irregular dark brown spots.

ENDIVE, ESCAROLE (see CHICORY)

GREENS

What To Look For: Leaves that are fresh, young, tender, free from blemishes and which have a good, healthy green color. Beet tops and ruby chard show reddish color.

What To Avoid: Leaves with coarse, fibrous stems, yellowish green color, softness (a sign of decay) or a wilted condition. Also avoid greens with evidence of insects—especially aphids—which are sometimes hard to see and equally hard to wash away.

KALE (see GREENS)

LETTUCE

What To Look For: Iceberg and Romaine lettuce leaves should be crisp. Other lettuce types will have a softer texture, but leaves should not be wilted. Look for a good, bright color, medium to light green in most varieties.

What To Avoid: Heads of iceberg type that are very hard and which lack green color (signs of over-maturity). Such heads sometimes develop discoloration in the center of the leaves (the "mid-ribs"), and may have a less attractive flavor. Also avoid heads with irregular shapes and hard bumps on top, which indicate the presence of overgrown central stems.

Check the lettuce for tip burn, a tan or brown area (dead tissue) around the margins of the leaves. Slight discoloration of the outer or wrapper leaves usually will not hurt the quality of the lettuce, but serious discoloration or soft decay definitely should be avoided.

MUSHROOMS

What To Look For: Young mushrooms that are small to medium in size. Caps should be either closed around the stem or moderately open with pink or light tan gills. The surface of the cap should be white or creamy or from some producing areas light brown.

What To Avoid: Overripe mushrooms (shown by wide-open caps and dark, discolored gills underneath) and those with pitted or seriously discolored caps.

OKRA

What To Look For: Tender pods (the tips will bend with very slight pressure) under 4^1/$_2$ inches (11 cm) long. They should have a bright green color and be free from blemishes.

What To Avoid: Tough, fibrous pods, indicated by tips which are stiff and resist bending, or by a very hard body of the pod, or by pale, faded green color.

ONIONS (DRY ONIONS)

What To Look For: Hard or firm onions that are dry and have small necks. They should be covered with papery outer scales and reasonably free from green sunburn spots and other blemishes.

What To Avoid: Onions with wet or very soft necks, which usually are immature or affected by decay. Also avoid onions with thick, hollow, woody centers in the neck or with fresh sprouts or green sunburn spots.

ONIONS (GREEN), SHALLOTS, SCALLIONS, LEEKS

What To Look For: Crisp, tender bunches with fresh green tops. Choose those with 2 – 3 inches (5 – 7 cm) of white above a slightly bulbed root end.

What To Avoid: Yellowing, wilted, discolored, or decayed tops (indicating flabby, tough or fibrous condition of the edible portions). Bruised tops will not affect the eating quality of the bulbs if the tops are removed.

PARSNIPS

What To Look For: Parsnips of small or medium width that are well formed, smooth, firm, and free from serious blemishes or decay.

What To Avoid: Large, coarse roots (which probably have woody, fibrous, or pithy centers), and badly wilted and flabby roots (which will be tough when cooked).

PEAS

What To Look For: Young, bright-green, angular pods, well filled with well-developed peas that snap readily.

What To Avoid: Yellow or whitish color indicates over mature and tough product. Mildew, swollen or speckled pods indicate poor quality.

PEPPERS (SWEET GREEN, RED AND YELLOW)

What To Look For: Medium to dark green color (or bright red or yellow), glossy sheen, relatively heavy weight, and firm walls or sides.

What To Avoid: Peppers with very thin walls (shown by light weight and flimsy sides), punctures through the walls, and peppers with soft watery spots on the sides (evidence of decay).

POTATOES

What To Look For: In new potatoes—well shaped, firm potatoes that are free from blemishes and sunburn (a green discoloration under the skin that should be removed before consumption). Some amount of skinned surface is normal, but potatoes with large skinned and discolored areas are undesirable. General all-purpose and baking potatoes—reasonably smooth, well shaped firm potatoes free from blemishes, sunburn, and decay. These potatoes should be free from skinned surfaces.

What To Avoid: Potatoes with large cuts or bruises (they will mean waste in peeling), those with a green color (probably caused by sunburn or exposure to light in store), and potatoes showing any signs of decay. Also avoid sprouted or shriveled potatoes.

RADISHES

What To Look For: Medium size radishes (³/₄ to 1¹/₈ inches or 2 to 3 cm in diameter) that are plump, round, firm, and of good red color.

What To Avoid: Very large or flabby radishes (likely to have pithy centers). Also avoid radishes with yellow or decayed tops.

RHUBARB

What To Look For: Fresh, firm rhubarb stems with a bright, glossy appearance. Stems should have a large amount of pink or red color, although many good quality stems will be predominantly light green. Be sure the stem is tender and not fibrous.

What To Avoid: Either very slender or extremely thick stems, which are likely to be tough; also avoid wilted or flabby rhubarb.

RUTABAGAS (*see* TURNIPS)

SPINACH (*see* GREENS)

SQUASH (SUMMER)

What To Look For: Squash that is tender but firm, well formed and developed, fairly heavy in relation to size and fresh appearing. The skin of a tender squash is glossy instead of dull, and it is neither hard nor tough. Seeds are soft and fully edible.

What To Avoid: Over mature squash, which will have a dull appearance and a hard, tough surface. Such squash usually have enlarged seeds and dry stringy flesh.

SQUASH (FALL AND WINTER)

What To Look For: Full maturity, indicated by a hard, tough rind. Also look for squash that is heavy for its size (meaning a thick wall and more edible flesh).

What To Avoid: Squash with cuts, punctures, sunken or moldy spots on the rind—all indications of decay. A tender rind indicates immaturity that is a sign of poor eating quality in winter squash varieties.

SWEET POTATOES

What To Look For: Well-shaped firm sweet potatoes with smooth, bright, uniformly colored skins, free from signs of decay. Because they are more perishable than white potatoes, extra care should be used in selecting sweet potatoes.

What To Avoid: Sweet potatoes with worm holes, cuts, grub injury, or any other defects which penetrate the skin; this causes waste and can readily lead to decay. Even when the decayed portion is cut away, the remainder of the potato flesh that looks normal may have a bad taste.

Decay is the worst problem with sweet potatoes and is of three types: wet, soft decay; dry firm decay which begins at the end of the potato, making it discolored and shriveled; and dry rot in the form of sunken, discolored areas on the sides of the potato.

TOMATOES

What To Look For: Tomatoes which are well formed, smooth, well-ripened and reasonably free from blemishes. For fully ripe fruit, look for an overall rich red color and a slight softness. Softness is easily detected by gentle handling. For tomatoes slightly less than fully ripe, look for firm texture and color ranging from pink to light red.

What To Avoid: Overripe and bruised tomatoes (they are both soft and watery) and tomatoes with sunburn (green or yellow areas near the stem scar) and growth cracks (deep cracks around the stem scar). Also avoid decayed tomatoes which will have soft, water-soaked spots, depressed areas, or surface mold.

TURNIPS (INCLUDING RUTABAGAS)

What To Look For—Turnips: Small or medium size, smooth, fairly round, and firm vegetables. If sold in bunches, the tops should be fresh and should have a good green color.

What To Avoid: Large turnips with too many leaf scars around the top and with flabby, soft or obvious fibrous roots.

What To Look For—Rutabagas: Heavy weight for their size, generally smooth, round or moderately elongated shape. Good quality rutabagas should also be firm to the touch.

What To Avoid: Rutabagas with hard texture, pithy, skin punctures, deep cuts, or decay.

WATERCRESS

What To Look For: Watercress that is fresh, crisp and bright-green color with rather long stems.

What To Avoid: Bunches with yellow, wilted, bruised or decayed leaves.

COOKING METHODS

FRESH VEGETABLES

Vegetables can be cooked by boiling, baking, microwaving, frying, steaming, stir-frying or pressure cooking. The length of time required for a given vegetable to cook by any method cannot be stated exactly. Cooking time differs with the variety and maturity of each vegetable, the period and the temperature at which the vegetable was held after it was harvested, and the size of the pieces. Each vegetable should be cooked for the shortest time necessary to give a palatable product. See timetable for cooking fresh vegetables, Table 8.1.

FROZEN VEGETABLES

The length of time required for cooking (boiling) frozen vegetables is usually less than that required for fresh vegetables. This is because the blanching and freezing of vegetables tenderize them to some degree. Frozen vegetables should be put into boiling water, brought quickly to a boil, then boiled gently until just tender. Because cooking time varies with different vegetables, package directions are the best guide for proper cooking of frozen vegetables.

USE OF PRESSURE SAUCEPAN

Manufacturer's directions for use of a pressure saucepan are the best guide because they are based on that specific model. Vegetables are usually cooked at 15 pounds (103.4 kPa) pressure. One note of caution: At the end of the cooking time, the pressure saucepan should be placed in cold water or under running cold water to reduce the pressure quickly and stop the cooking.

USE OF MICROWAVE

Manufacturer's directions for use with a microwave are the best guide since they are based on that specific model. The Timetable for Cooking Vegetables suggests approximately a three-minute resting time following the recommended cooking time. Food continues to cook during this period.

STORAGE AND USE

- Store the following vegetables in refrigerator crisper or in a plastic bag in refrigerator: asparagus, beets, broccoli, Brussel sprouts, beet greens, cabbage, carrots, cauliflower, celery, chard, collard greens, cucumber, green onions, lettuce, mushrooms, mustard greens, okra, peppers, spinach, summer squash, and turnip greens.

- Store sweet corn in husks, uncovered in refrigerator.

- Store ripe tomatoes, uncovered, in the refrigerator. Keep unripe tomatoes at room temperature, but away from direct sunlight until they ripen.

- Store dry onions in loosely woven or open mesh containers at room temperature or slightly cooler temperatures.

- Store potatoes in a dark, dry, well ventilated place at temperatures between 45° and 50° F (7° and 10° C). Potatoes stored at room temperature need to be used sooner than those stored under ideal conditions to prevent greening or sprouting and shriveling.

- Store these vegetables at 60°F (15°C); hard rind squashes, eggplant, rutabagas and sweet potatoes. Temperatures below 50° F (10° C) may cause chilling injury. If these vegetables must be kept at room temperature, plan to use them within a week.

- Use these vegetables within one or two days: asparagus, beet greens, broccoli, Brussel sprouts, chard, collard greens, green peas, green onions, kale, lettuce and other salad greens, fresh lima beans, mushrooms, mustard greens, spinach, sweet corn and turnip greens.

TABLE 8.1

Fresh Vegetables Cooking Timetable

Vegetable	Boiling*	Steaming*	Microwave (700 Watts)	Baking
Artichokes				
French or globe, whole	35-45 minutes		High 4-5 minutes	
Jerusalem whole	25-35 minutes	35 minutes	1 medium†	30-60 minutes
Asparagus				
Whole or butts	10-20 minutes	12-30 minutes	Spears—	
Tips	5-15 minutes	7-14 minutes	High 6-9 min./lb. 500g	
			Cuts—High 5-7 min†	
Beans				
Lima/green	25-30 minutes	25-35 minutes	Green beans—	
Soy, green	20-30 minutes	25-35 minutes	High 6-10 min. per ½ lb./	
Green, whole or			250 mg	
1″ pieces	15-30 minutes	20-35 minutes		
French	10-20 minutes	15-25 minutes	High 5-7 min. per lb./500 g†	
Broccoli				
Heavy stalk, split	10-15 minutes	15-20 minutes	Spears—High 8-12 min. per 1½ lb./500-750 g†	
Brussel sprouts				
Whole	10-20 minutes	15-20 minutes	High 4-8 min. per lb. (4 cups)/ 500 g (1 l)†	
Cabbage				
Green			Green, red or savory	
quartered	10-15 minutes	15 minutes	wedges—High 7½-13½ min. per lb./500 g†	
shredded	3-10 minutes	8-12 minutes	7½ -13½ min. per lb./500 g	
Red				
shredded	8-12 minutes	8-12 minutes	7-13 min. per lb./500 g†	
Carrots				
Young				
whole	15-20 minutes	20-30 minutes	High 6-8 min. per 12 oz./	35-40 minutes
sliced	10-20 minutes	15-25 minutes	340 g†	30-40 minutes
Mature			Cut (2″) or 10-12 med.	
whole	20-30 minutes	40-50 minutes	High 6-8 minutes	60 minutes
sliced	15-25 minutes	25-30 minutes	Slices (¼″)—High 4-7 min. per lb./500 g†	

* For altitude cookery, increase cooking time 4-11 percent at 5,000 feet (1524 m); 20-25 percent at 7,200 ft (2194 m); and 55-66 percent at 10,000 feet (3048 m) and add more water.

† Rest covered 3 minutes.

TABLE 8.1 (CONTINUED)
Fresh Vegetables Cooking Timetable

Vegetable	Boiling*	Steaming*	Microwave (700 Watts)	Baking
Cauliflower				
Whole	15-25 minutes	25-30 minutes	High 5½-7½ min. per lb./500 g†	
Flowerets	8-15 minutes	10-20 minutes	High 4-7 min. per lb./500 g†	
Celery				
Diced	15-18 minutes	25-30 minutes	Slices (¼") High 5-8 min. per lb./500 g†	
Chard				
Swiss	10-20 minutes	15-25 minutes	High 5½ - 6½ min. lb./500 g†	
Collards	10-20 minutes		High 5½ - 6½ min. per lb./500 g†	
Corn				
Kernel	10-12 minutes	10-12 minutes	High 6-7 min. per lb./500 g†	
On cob	6-12 minutes	10-15 minutes	Husked—High 2-5 min. for one ear; 4½-10 min. for two ears†	
Eggplant				
Sliced	10-20 minutes	15-20 minutes	High 7-10 min. per lb./500 g†	
Kale	10-15 minutes		High 6-7 min. per lb./500 g†	
Kohlrabi				
Slices	20-25 minutes	30 minutes	High 10-15 min. per 4-5 med (2 lb./1 kg)†	
Okra				
Sliced	10-15 minutes	20 minutes	High 7-10 min. per lb./500 g†	
Onions				
Small				
Whole	15-30 minutes	25-35 minutes	High 7-10 min. (4 medium)†	
Large				
Whole	20-40 minutes	35-40 minutes		50-60 minutes
Parsnips				
Whole	20-40 minutes	30-45 minutes	High 7-8 min. per lb./500 g†	45-60 minutes
Quartered	8-15 minutes	30-40 minutes		

* For altitude cookery, increase cooking time 4-11 percent at 5,000 feet (1524 m); 20-25 percent at 7,200 ft (2194 m); and 55-66 percent at 10,000 feet (3048 m) and add more water.

† Rest covered 3 minutes.

TABLE 8.1 (CONTINUED)
Fresh Vegetables Cooking Timetable

Vegetable	Boiling*	Steaming*	Microwave (700 Watts)	Baking
Peas				
Green	12-16 minutes	10-20 minutes	High 5-7 min. per lb. (2 cups)/ 500 g (500 ml) Pea Pod—High 2-4 min. 1/4 lb./125 g†	
Potatoes				
White				
Medium	25-40 minutes	30-45 minutes	Baked—High 3-5 min. one	45-60 minutes
Whole			med; 5-7½ min. two med† Boiled whole—High 5-8	
Quartered	20-25 minutes	20-30 minutes	min. per eight med Slices—(1/4") 5-7 min. per 4 med†	
Rutabaga				
Diced	20-30 minutes	35-40 minutes	High 14-18 min. per 1½ lbs. (3-4 cups)/750 g (0.7-1 l)†	
Spinach	3-10 minutes	5-12 minutes	High 7½ min. per lb./500 g†	
Squash				
Hubbard				
2" pieces	15-20 minutes	25-40 minutes	Winter—High 6-7 min. per	40-60 minutes
Summer			lb./500 g†	
Sliced	8-15 minutes	15-20 minutes	Summer—(1/4" slices) High 2½-6½ min. per 2 cups/ 500 ml†	30 minutes
Sweet Potatoes				
Yams				
Whole	30-35 minutes	35-55 minutes	High 3-5 min. per potato;	30-45 minutes
Quartered	15-25 minutes	25-30 minutes	5-9 min. for two potatoes	
Tomatoes	7-15 minutes		High 4-6 min. per lb./500 g†	15-30 minutes
Turnips				
Whole	20-30 minutes		Slices (1/4") High 9-11 min.	
Sliced	15-20 minutes	20-25 minutes	4 med Cubes (1/2") High 12-14 min. per 4 med†	

* For altitude cookery, increase cooking time 4-11 percent at 5,000 feet (1524 m); 20-25 percent at 7,200 ft (2194 m); and 55-66 percent at 10,000 feet (3048 m) and more water added.

† Rest covered 3 minutes.

TABLE 8.2
Guide to Herb-Vegetable Cookery

Vegetable	Appropriate Spice or Herb	Vegetable	Appropriate Spice or Herb
Asparagus	Cayenne or red pepper, celery flakes, cinnamon, mustard seed, nutmeg, poppy seeds, sesame seed or tarragon	**Mushrooms**	Herb seasoning, marjoram or sesame seed
		Okra	Celery seed
Lima beans	Marjoram, oregano, rosemary, sage, savory, tarragon or thyme	**Onions**	Caraway seed, cayenne or red pepper, chili powder, cloves, mustard seed, nutmeg, oregano, parsley flakes, poultry seasoning, sage or thyme
Green beans	Basil, dill, marjoram, mint, mustard seed, nutmeg, oregano, poultry seasoning, rosemary, savory, tarragon or thyme	**Peas**	Basil leaves, dill, marjoram, mint, oregano, poppy seed, rosemary, sage or savory
Beets	Allspice, bay leaves, caraway seed, cloves, cinnamon, dill, ginger, mustard seed, savory or thyme	**Potatoes**	Basil leaves, bay leaves, caraway seed, celery seed, chives, dill, herb seasoning, mustard seed, oregano, parsley flakes, poppy seed, rosemary or thyme
Broccoli	Caraway seed, cayenne or red pepper, celery flakes, dill, herb seasoning, lemon or orange bits, mustard seed, nutmeg, tarragon or thyme		
		Rutabagas	Pumpkin pie spice
Brussel sprouts	Basil, caraway seed, dill, mustard seed, rosemary, sage or thyme	**Spinach**	Allspice, basil leaves, lemon or orange bits, mace, marjoram, nutmeg or oregano
Cabbage	Caraway seed, celery seed, dill, mint, mustard seed, nutmeg, rosemary, savory or tarragon	**Squash**	Allspice, basil leaves, cardamom, cinnamon, cloves, fennel, ginger, lemon or orange bits, marjoram, mustard seed, nutmeg, oregano, pumpkin pie spice, rosemary or thyme
Carrots	Allspice, bay leaves, basil leaves, caraway seed, chili powder, cinnamon, dill, fennel, ginger, lemon or orange bits, mace, marjoram, mint, nutmeg, oregano, parsley flakes, poppy seed, pumpkin pie spice, rosemary or thyme		
		Sweet potatoes	Allspice, cardamom, cinnamon, cloves, ginger, lemon or orange bits, or nutmeg
		Tomatoes	Basil, bay leaves, celery seed, cloves, herb seasoning, oregano, parsley flakes, rosemary, sage, sesame seed, tarragon or thyme
Cauliflower	Basil leaves, caraway seed, celery salt, dill, mace, mustard seeds, parsley flakes, rosemary or tarragon		
Corn	Cayenne or red pepper, celery flakes, chili powder or mustard seeds	**Turnips**	Oregano, poppy seed or rosemary
Cucumbers	Basil, dill, mint or tarragon	**Green salads**	Basil, chives, dill or tarragon
Eggplant	Allspice, cayenne or red pepper, chili powder, basil leaves, herb seasoning, marjoram, oregano, parsley flakes, poultry seasoning or sage		

Note: Curry powder adds piquancy to creamed vegetables. Pepper and parsley may be used with any of the vegetables.

Based on "Spices and Herbs, Vegetables in Family Meals," *Home and Garden Bulletin No. 105*, U.S. Department of Agriculture, 1965, and "Cooking Magic with Herbs and Spices," McCormick and Co., Inc.

Grain Products

Cereal grains are the seeds of cultivated grasses, including barley, corn, oats, rice, and wheat. The kernels of the various grains are similar in structure, but differ in size and shape. The three main structural parts of the kernel are bran, which is the outer protective covering of the kernel; endosperm, which comprises about 85 percent of the kernel and contains the food supply of the plant; and germ, which contains elements necessary for new plant life.

BARLEY PRODUCTS

In the United States, this grain is sold mainly as pearl barley that is the whole grain with hull and bran removed. It is used principally as a soup ingredient. Pearl barley may be used in casseroles, pilafs, or salads. The grain may also be made into a flour by a process similar to that for making wheat flour. Barley flour and milling fractions such as bran, middlings, and shorts are used in baked products and breakfast cereals. Barley products are high in soluble fiber.

CORN PRODUCTS

Cornmeal is made by grinding white or yellow corn kernels to a fineness specified by federal standards. Cornmeals contain small amounts of fat and crude fiber and not more than 15 percent moisture. Degermed cornmeal is made by grinding the kernel after the germ is removed.

Bolted white or yellow cornmeal is ground finer than the above type but otherwise is similar.

Enriched cornmeal contains added vitamins and minerals. It is enriched with thiamin, riboflavin, niacin, and iron. It may also contain calcium and Vitamin D.

Corn flour may be a byproduct in the preparation of cornmeal or may be prepared especially by milling and sifting yellow or white corn.

Corn grits, grits, or hominy grits are made from white or yellow corn from which the bran and germ have been removed. Grits are more coarsely ground than cornmeal.

Hominy is corn with the hull and germ removed, left whole, or broken into particles. Pearl hominy is whole grain hominy with the hulls removed by machinery. Lye hominy is whole grain hominy that has been soaked in lye water to remove the hulls. Granulated hominy is a ground form of hominy. Hominy grits are broken grains.

Cornstarch is the refined starch obtained from the endosperm of corn.

Waxy cornstarch is prepared from waxy corn. It is composed almost completely of amylopectin with little or no amylose. This starch acts as a stabilizer for frozen sauces and pie fillings.

Flavored cornstarch mixes are blends of cornstarch, sugar, and flavorings for making puddings and pie fillings.

Instant cornstarch puddings usually contain dehydrated gelatinized cornstarch, sugar or sugar substitute, and flavoring.

Corn cereals are usually ready-to-eat flakes or puffs, made from corn grits that have been cooked and dried or toasted. Corn cereals may be flavored, sugarcoated, and enriched with thiamin, riboflavin, niacin, and iron, or fortified with vitamins and minerals.

OAT PRODUCTS

Oatmeal, or rolled oats, also called oats, is made by rolling the groats (oats with hulls removed) to form flakes. Regular oats and quick-cooking oats differ only in thinness of flakes. For quick-cooking oats, the finished groats (edible portion of the kernel) are cut into tiny particles that are then rolled into thin, small flakes. Steel-cut oats, those that are not flaked, are the most resistant of the oat products to overcooking. Although oatmeal is primarily considered a breakfast cereal, many recipes call for its use in cooking and baking.

Oat bran is essentially the envelope of the groat, composed of the outermost pericarp. Oat bran is high in soluble fiber and can be used in many types of recipes.

RICE PRODUCTS

Rice grains are classified as long grain, medium grain, and short grain, according to varieties. Medium and short grain rice have a higher proportion of amylopectin than long grain rice and therefore the grains cling together after cooking. **Brown rice** is the grain from which only the hull has been removed. It has a nutty flavor when cooked. The bran and germ are high in fiber, minerals, and vitamins. **White rice** has the bran and germ layers removed and is the starchy endosperm of the grain.

Aromatic rice has a natural aroma when cooked. Special varieties of rice are aromatic; nothing is added to the grains. Basmati rice and jasmine rice are imported aromatic rices. Aromatic rice is also grown in the U.S.

Enriched rice contains thiamin, riboflavin, niacin, folic acid, and iron. Over 90 percent of milled rice in the U.S. is enriched. Enriched rice should not be rinsed before or after cooking.

Imported rice includes arborio rice from Italy which is used to make risotto, a creamy rice dish. U.S. medium or short grain rice may be substituted for arborio rice.

Precooked or instant rice is packaged long grain rice (brown or white) that has been cooked, rinsed, and dried by a patented process.

Parboiled or converted rice has been steeped in warm or hot water, drained, steamed under pressure, and dried before it is hulled and milled. This steaming process causes vitamins and minerals present in the outer coats to migrate to the interior of the kernel. Parboiling gelatinizes some of the starch, which results in a shorter cooking time.

Rice bran consists of bran and germ. It is a smooth brown powder with a faintly sweet taste; it is rich in nutrients and fiber.

Rice polish consists of inner bran layers and some endosperm. It is a smooth yellow powder with a sweet taste.

Rice flour is made from white or brown rice ground into flour. It is gluten-free and can be used by those who are allergic to wheat flour.

Waxy rice flour is made from waxy rice. It is composed almost completely of amylopectin with little or no amylose. It acts as a stabilizer in sauces and gravies and is especially useful in preventing separation in these products when they are frozen.

CLASSES OF WHEAT

There are six classes of wheat grown in the United States. They are durum wheat for pasta; hard red spring, hard red winter, and hard white for yeast breads and hard rolls; soft red winter for flat breads, cakes, pastries, and crackers; soft white wheat for flat breads, cakes, pastries, crackers, and noodles.

WHEAT PRODUCTS

Bulgur wheat is whole wheat that has been cooked, dried, partly debranned, and cracked into coarse, angular fragments. Rehydration requires simmering for 15 to 25 minutes. It may be used as an alternate for rice in many recipes and resembles whole wheat in nutritive properties.

Cracked wheat is similar to bulgur in nutrition and texture, but it has not been precooked. It is the whole kernel broken into smaller pieces which serves a hot cereal or an addition to baked goods. Cracked wheat should be presoaked or cooked before addition to baked products.

Farina is the coarsely ground endosperm of wheats other than durum. It is the prime ingredient in many hot breakfast cereals and is occasionally used for pasta. Enriched farina contains thiamin, riboflavin, niacin, folic acid, and iron. Calcium and Vitamin D may also be added.

Semolina is the coarsely ground endosperm of durum wheat. The primary use of semolina is for pasta; however, it is also used for couscous, a tiny rice-grain sized pasta used in Moroccan cooking.

Wheat bran is the outer layer of the kernel. It is high in insoluble fiber and can be added to baked goods, casseroles, cereals, and meat dishes.

Wheat germ is the germ of the kernel and is often added

to baked goods, casseroles, and beverages to enhance the nutritional value and give a nutty, crunchy texture. Because of the oil content, it should be tightly covered and stored in the refrigerator or freezer.

Wheat kernels (berries) are the whole-grain kernel and can be sprouted and added to salads and baked goods. The cooked berries can be used as a meat extender, breakfast cereal, or a substitute for beans in chili, salads, and baked dishes.

FLOURS

The term *flour* when used in recipes is understood to mean refined all-purpose wheat flour unless otherwise designated as, for example, bread flour, cake flour, self-rising flour, oat flour, or rice flour. The majority of states require white flour to be enriched. Therefore, the major mills in the United States enrich all their flour so it can be shipped interstate. The Federal standards for enriched corn, rice, macaroni, and wheat flour are presented in Table 9.1. The following describes in more detail the various types of wheat flour available in most markets.

All-purpose flour is the finely ground endosperm of the wheat kernel separated from the bran and germ during milling. It is made from hard wheats (other than durum) or a combination of hard and soft wheats. It is best for quick breads and may be used for yeast breads, cakes, pastry, and cookies.

Bread flour is the ground endosperm from hard wheats and is best for yeast breads because of its high gluten content (higher in percent and stronger in viscoelastic qualities than all-purpose flour).

Cake flour is milled from the endosperm of soft wheats and is especially suited for cakes, cookies, and crackers. Because it is so fine and uniform, it feels soft and satiny. It is low in gluten and forms a more delicate structure than all-purpose flour.

Enriched flour is white flour which contains added vitamins and minerals. Thiamin, riboflavin, niacin, folic acid, and iron are added; the addition of calcium is optional.

Gluten flour is a mixture of wheat flour and gluten with a protein content of about 45 percent vs. 13 percent for bread flour, 11 percent for all-purpose, and 8-9 percent

TABLE 9.1

Federal Standards for Enriched Corn, Rice, Macaroni and Wheat Flour

Nutrient	Farina	Corn Meal	Rice	Macaroni	Wheat
Required per pound (454 g) of flour:					
Thiamin	2.0–2.5 mg	2.0–3.0 mg	2.0–4.0 mg	4.0–5.0 mg	2.9 mg
Riboflavin	1.2–1.5 mg	1.2–1.8 mg		1.7–2.2 mg	1.8 mg
Niacin	16–20 mg	16–24 mg	16–32 mg	27–34 mg	24 mg
Iron	13 mg	13–26 mg	13–26 mg	13–16.5 mg	20 mg
Folic Acid	0.7–0.87 mg	0.7–1.0 mg	0.7–1.4 mg	0.9–1.2 mg	0.7 mg
Optional per pound (454 g) of flour:					
Calcium	500 mg	500–750 mg		500–625 mg	960 mg
Vitamin D	250 IU	250–1000 IU		250–1000 mg	
Riboflavin			1.2–2.4 mg		

for cake flour.

Instant, instantized, instant-blending or quick mixing flour is a granular all-purpose flour which blends more readily with liquids than regular flour. It is made by exposing all-purpose flour to hot water or steam to combine the individual particles into agglomerates. It is excellent for gravies and as a thickener, but can also be used in baked goods with some adaptations.

Pastry flour is milled from the endosperm of soft wheats and is used in pastries.

Self-rising flour is an all-purpose flour with 1½ teaspoons (7.5 ml) of baking powder and ½ teaspoon (2.5 ml) of salt added per cup (250 ml) of flour. It can be substituted for an all-purpose flour by reducing the baking powder and salt in the recipe accordingly. Self-rising flour is used primarily in the southeastern U.S. for biscuits, cornbread, waffles, and shortbread, using special recipes.

Vital wheat gluten is used by bakers to enhance the gluten quality in flour or bread products containing fiber. It contains about 75 percent protein and is made by gently washing a flour-water dough which separates the gluten portion from the starch. The protein fraction is then dried to form a powder. Very little vital wheat gluten is needed to enhance a recipe. Use 1 teaspoon (5 ml) of gluten for every 6 cups (1.4 l) of flour and add an additional 1½ teaspoons (7 ml) water.

White-wheat flour is made from hard white wheat and is becoming popular in the U.S. When the whole kernel is ground, it produces a flour with the fiber and nutritional content of the darker whole-wheat flour, but with a lighter, golden color and less bitter flavor. It has excellent baking qualities but is not readily available nationally in grocery stores.

Whole-wheat flour, also called graham flour, is coarse in texture and is ground from the entire wheat kernel. Bran inhibits the gluten development and absorbs water, giving a heavier, denser loaf of bread. Many recipes call for a mixture of whole-wheat and white flour. Whole-wheat flour contains more dietary fiber than white flour, but the other nutrients are very similar to white flour.

PASTA

Pasta includes macaroni, noodles, and spaghetti in one of 150 shapes. Durum wheat is used to make semolina, which is enriched with thiamin, riboflavin, niacin, and iron; pasta dough is made from semolina and water. Noodles are made from a more finely ground semolina and have eggs added. Pasta dough is kneaded and then forced through dies to create the many different forms of pasta available. The pasta is dried by automatic dryers which remove moisture under carefully controlled conditions. Fresh pasta is available in the refrigerated section of the grocery store. Pasta should be cooked until "al dente," an Italian term meaning tender but firm. It is best to follow package directions, but keep in mind the following: use at least one quart (1 l) of water for every four ounces (115 g) of dry pasta; add pasta to boiling water; stir the pasta when you add it, when the water returns to a boil, and occasionally during cooking; drain in a colander; and undercook slightly if the pasta will be used as part of a dish that requires further cooking.

OTHER FLOURS, CEREALS, GRAINS, AND SEEDS

Amaranth is a tall willowy plant similar in height to corn. It has a large shaggy head containing thousands of tiny seeds. The seed can be milled into a whole grain flour, or puffed like rice or corn.

Buckwheat is neither a wheat nor a cereal grain, but the seed from any of the *Fagopyrum* family of herbs. *Kasha* is a word commonly used to describe cooked buckwheat groats, a Russian dish. Buckwheat is ground and used for its unique flavor to replace part of the wheat flour in buckwheat pancake mixes. It has more starch and less protein than wheat.

Millet has a stronger flavor than most cereals. It can be eaten as the whole grain. Millet can be husked, soaked, boiled, and ground into meal. It can be added in bread making.

Potato flour is prepared from cooked potatoes that have been dried and ground.

Quinoa (pronounced keenwah) is a small, disk-shaped seed that can come in many colors. It is not a cereal grain; it is the fruit of an annual herb. It has been grown in the mountain regions of Peru and Bolivia for 3,000 years. The whole grain is cooked like rice, and quinoa can also be ground into flour and used for baked goods.

Rye flour is the finely ground product obtained by sifting rye meal. It is available in three grades: white, medium, and dark. Wheat flour is the only flour containing gluten-forming proteins that are so important in baked products because they are both viscous and elastic. Rye flour produces gluten of low elasticity; often it is used with wheat flour for volume.

Sorghum is a tropical grass similar to Indian corn and has an unusual flavor that can be improved by steaming or dry heat processing. Small amounts of sorghum are sometimes added to crackers and snack foods. Sorghum is also used to make a syrup.

Soy flour is highly flavored. It must be combined with wheat flour when used in baked products, because it lacks gluten-forming proteins. The amount of liquid in a recipe must be increased when soy flour is substituted for part of the wheat flour. Full-fat soy flour is made by grinding soybeans that have only the hull removed. Low-fat soy flour is made from the press cake after all or nearly all of the oil is taken out of the soybeans. The soybeans may be heat treated or conditioned with steam prior to oil extraction.

Soy grits are made from coarsely ground soy press cakes and are a low-fat product.

Spelt (*triticum spela*) is a predecessor of wheat (*triticum sativum*) and has a very strong hull that is difficult to remove when milling. It is available primarily as flour, but lacks the ability of wheat flour to produce a top quality loaf.

Teff is the smallest grain in the world; 150 grains weigh the same as one grain of wheat. It is available as the whole grain and flour. Teff flour can be used with other flours in baked goods and as a thickener in gravy, soups, and stews.

Triticale is a hybrid cereal derived from crossing wheat and rye. It can be cooked as a whole grain, ground into flour for use in baking, or processed into cereal flakes.

Wild rice is the long brownish grain of a reed-like water

plant rather than a true rice. It is hulled but not milled.

BREAKFAST CEREALS

Whole-grain cereals retain the natural proportion of bran, germ, and endosperm, and the specific nutrients that are normally contained in the whole unprocessed grain.

Enriched cereals contain added amounts of thiamin, riboflavin, niacin, and iron. The levels of nutrient enrichment are established by FDA.

Fortified cereals contain added amounts of selected nutrients that may or may not have been present in the grain before processing.

Restored cereals contain added amounts of selected nutrients to supply the same approximate levels of these nutrients in the finished grain products as were present in the whole grain before processing.

Varieties of hot and ready-to-eat cereals include extruded, flaked, granulated, puffed, rolled, and shredded. Hot cereals include regular, quick-cooking, and instant varieties. Instant hot cereals are prepared for precooked dried grains. Disodium phosphate may be added, giving a high sodium content. The kernels may be modified with a small amount of an enzyme preparation to permit quick entry of water into the starch granules. Some ready-to-eat cereals are presweetened.

STORAGE AND MEASUREMENT OF GRAINS AND CEREAL PRODUCTS

■ Store flours, grains, and cereals in tightly covered containers to keep out dust, moisture, and insects; store in a dry place at room temperature. Whole grains should be stored in the refrigerator or freezer due to their oil content, which could oxidize and become rancid.

■ Store dried pasta or egg noodles tightly covered or well wrapped. These products may be stored for one year. Follow package directions for storage of commercially prepared fresh pasta. Fresh pasta must be refrigerated and should be eaten within two days or frozen for later use.

■ Cereals may be stored satisfactorily for two to three

months; corn meal and hominy grits for four to six months; bulgur and brown or wild rice for six months; and other rices for one year.

- To measure sifted or unsifted all-purpose flour, sift into a bowl. Spoon rounded tablespoons of the flour lightly into a fractional measuring cup until the flour overflows the cup. Level with the straight edge of a spatula or knife. Sifting flour results in consistent, accurate measures, even if presifted when produced. Although some believe air is incorporated by sifting, it is more likely that the measure of flour is slightly lower. Therefore, it is suggested in the Substitutions section that if flour is not sifted at home, 2 tablespoons should be removed from a cup measurement.

- To measure whole-grain flours, instant flour, and meals, stir lightly with a fork or spoon but do not sift before measuring or the particles could be removed. Follow measuring directions for all-purpose flour.

The dietary fiber of grains and cereal products are shown in Table 9.2. Cereal sources of various fiber components are as follows:

Cellulose
  Whole wheat flour
  Bran

Gums
  Oat products
  Barley products

Hemicellulose
  Bran
  Cereals
  Whole grains

Lignin
  Whole wheat flour

TABLE 9.2

Dietary Fiber Content of Grains and Cereal Products
(in commonly served portions)

Fiber Categories	<1	1 – 1.9 g	2 – 2.9 g	3 – 3.9 g
Foods				
Cereals (1 oz., 28 g)	There are cereals in every fiber category. Check labels.			
Pasta (1 cup, 250 ml)		macaroni spaghetti		whole wheat spaghetti
Rice (½ cup, 125 ml)	white	brown		

Sweetening Agents, Fats and Oils, Leavening Agents

SWEETENING AGENTS

SUGARS

The term sugar, when unqualified as to source, refers to refined sucrose derived from sugar beets or sugarcane. Except for source, these two sugars are the same and are 99.9 percent pure sucrose.

White Sugar

Granulated sugar is the standard product for general use, variously branded "granulated," "fine granulated," or "ultra-fine granulated." The variation indicates preference in terminology of the manufacturer rather than any definite particle size. It is available in numerous sizes and types of packages, from one-pound cartons to 100-pound bags.

Ultra-fine granulated sugar is a specially screened, uniformly fine-grained sugar designed for special use in cakes and in mixed drinks and other uses where quick creaming or rapid dissolving is desirable. It is available in one-pound cartons.

Powdered or confectioner's sugar is granulated sugar crushed and screened to a desired fineness. These are available in several degrees of fineness designated by name or number of ×'s following the name. Although 10× is finer than 4×, no standard terminology applies to all brands. It is used in frostings and icings and for dusting on baked products. It usually contains three percent cornstarch to prevent caking.

Special Forms of Sugar

Cut tablets are made from sugar that is molded into slabs that are afterward cut or clipped. Cut tablets of various sizes and shapes are packed in one- and two-pound cartons.

Pressed tablets are made by compressing moist, white sugar into molds to form the tablets which are afterward oven-dried to produce hard, smooth tablets. Tablets of various sizes and shapes come in one- and two-pound cartons.

Cubes, like pressed tablets, are formed in molds. Sizes range from 200 to 80 pieces to a pound. Cubes are packed in one- and two-pound cartons.

Brown Sugar

Brown sugar is a product that contains varying quantities of molasses, non-sugars (ash) naturally present in molasses, and moisture. It may be produced from the syrup remaining after the removal of commercially extractable white sugar or by the addition of refined syrups to specially graded, uniformly minute white sugar crystals. Four grades are available for food manufacturing but only two for consumer purchase. These are variously designated as "golden brown" or "light brown" and "dark" or "old-fashioned brown," indicating the color characteristic. Intensity of molasses flavor increases with color. The granulated (Brownulated®) form of brown sugar contains enough molasses to provide a flavor of an intensity between light and dark brown sugars; its use in baking requires adjustments in amounts of ingredients. Brown sugar imparts flavor and color to candies, condiments, baked goods, and the like. It is packed in 14 oz. and one-pound cartons and two-pound plastic bags. Dark brown sugar is more acidic than light brown sugar, and the proportions of light to dark affects the crystallization of candies.

Other Sugars

Raw sugar is the product crystallized at the stage in cane-sugar manufacture at which sugar goes from the mill to the refiner. Raw sugar may contain molds, fibers, waxes, and other contaminants.

Turbinado sugar is a partially refined sugar similar in appearance to raw sugar, which is sold for consumption without further refining.

Blended sugar is a combination of sugar and dextrose that may be used in place of an equal quantity of sugar, although its properties are somewhat different.

Corn sugar is dextrose (glucose) obtained by treating cornstarch with acid.

Fructose (levulose) is one of two sugars (the other is glucose) that results when sucrose reacts chemically with water. It may be produced industrially by the use of an enzyme (isomerase) to change the structure of glucose to yield fructose. It is highly soluble and is approximately one and one-half times as sweet as sucrose in sugar solutions.

Maple sugar is the solid product resulting from the evaporation of maple sap or maple syrup. It consists mostly of sucrose with some invert sugar and ash.

SYRUPS

From Sugarcane

Cane syrup is the concentrated sap of sugar cane. It is made by evaporation of the juice of sugarcane or by solution of sugar cane concrete (concentrate). The recommended maximum ash content of the unsulfured product is 4.5 percent; sulfured, 6 percent.

Molasses is the mother liquid from which raw cane sugar has crystallized. The following types are usually found:

Table molasses, which is light in color, contains a higher percentage of sugar and a smaller percentage of ash than are present in cooking molasses. It is often sold as light molasses.

Cooking (blackstrap) molasses is darker in color and has a stronger flavor than light molasses. Barbados molasses, which is specially treated cooking molasses, resembles cane syrup more than molasses in composition. It is often sold as dark molasses.

Refiners' syrup is the residual product obtained in the process of refining raw cane sugar that has been subjected to clarification and decolorization. It is a solution of sucrose and partially inverted sucrose, containing not more than 28 percent moisture. It is used extensively for flavoring corn syrup.

From Sorghum Cane

Sorghum syrup is obtained by concentration of the juice of the sugar sorghum plant. It contains not more than 30 percent water nor more than 6.25 percent ash calculated on a dry basis.

From Corn

Corn syrup (unmixed) is obtained by partial hydrolysis of cornstarch by use of acid or enzymatic catalysts or a combination of these. The resulting liquid is neutralized, clarified, and concentrated to syrup consistency. The principal ingredients are dextrose (glucose), maltose, and dextrins. Two types are commonly marketed.

Light corn syrup is corn syrup that has been clarified and decolorized. Retail products may have high fructose corn syrup (HFCS) and/or flavoring added.

Dark corn syrup is a mixture of corn syrup and refiners' syrup. It is used as a table syrup and also for the same purposes as light corn syrup in combinations that give a desirable darker color and distinctive flavor. Color and flavors may be added to retail products.

A third form available to food processors is **high fructose corn syrup (HFCS)**. An enzymatic process is used to produce syrup with 42 percent, 55 percent, or 90 percent fructose. The more fructose, the sweeter the syrup will be.

From Maple Trees

Maple syrup is made by evaporation of maple sap or by solution of maple sugar. It contains not more than 35 percent water and weighs not less than 11 pounds to the gallon.

From Bees

Honey is the nectar of plants, gathered, modified, stored, and concentrated by honeybees. The water content of honey is limited to about 20 percent. Its principal ingredients are levulose (fructose) and dextrose (glucose). The term honey in cookery refers to honey extracted from the comb. The different flavors of honey are classified according to the plant from which the nectar is derived.

By Special Processes

Blended syrups are mixtures of different, but somewhat similar, types of syrups that are sold for table purposes.

The composition of blended syrups is stated on the label.

Spray-dried syrups consist essentially of the solids of syrup which have been spray-dried.

STORAGE AND USE

- To store honey and syrups, keep the unopened containers at room temperature. Once the containers have been opened, refrigerate honey and syrups to protect against mold. If crystals form, place the container in hot water.

- Store white granulated sugar, covered, in a dry place. If the sugar becomes lumpy, sift before measuring.

- Store brown sugar in a plastic bag in airtight container. If sugar hardens, place a piece of foil or plastic wrap directly on the sugar and set a wad of dampened paper towel on the foil. Cover container tightly. The sugar will absorb the moisture and become soft. Remove paper when it has dried out. Brown sugar may be kept in the freezer to prevent drying.

- To soften brown sugar quickly, warm in a 250°F oven (120°C) and measure as soon as the sugar becomes soft. Sugar may also be heated in an open microwavable container in the microwave unit with a second container of $1/2$ cup (125 ml) water for 1–3 minutes (depending on quality and unit). It will harden again upon cooling.

- When measuring brown sugar, pack it firmly enough into the measuring cup for the sugar to retain the shape of the cup when turned out.

- Store powdered sugar in an airtight container to keep out moisture. If sugar becomes lumpy, sift before measuring.

NO-CALORIE AND LOW-CALORIE SWEETENERS

No-Calorie and Low-Calorie Sweeteners are sugar substitutes that provide sweetness but cause little change in viscosity or density of solutions and lack other functional properties of sugar. Because of the intensity of sweetness, very small amounts need to be used and calorie contributions, even of those that are metabolized, are considered to be insignificant. For best flavor, several are frequently combined and dextrin, lactose, or other bulking material included in the product. Generally a one-serving package equals the sweetness of two teaspoons sugar.

Currently, sweeteners which have been approved by the Food and Drug Administration or for which petitions for approval have been submitted are shown in Table 10.1.

RELATED PRODUCTS

Polydextrose is a condensation polymer of dextrose, containing minor amounts of bound sorbitol and citric acid. It has one calorie per gram. As a bulking agent, it gives texture and body to foods when used partially to replace sugar or fat in some foods. Since it has no sweetness, sweeteners may also be needed.

Polyols (polyhydric alcohols). Sugar alcohols, such as sorbitol, mannitol, and xylitol have limited use in foods. Although less sweet than sugar, they contribute to texture.

Tables 10.2 and 10.3 on pages 129 and 130 present information on Solubility of Sugar and Temperatures and Tests for Syrup and Candies, respectively.

FATS AND OILS

Fats and oils are substances of plant and animal origin that belong to a class of chemical compounds called lipids. The terms commonly used are "fats" that are lipids solid at room temperature, while "oils" are lipids that are liquid at room temperature. Most of the lipids found in foods are triglycerides, which are esters of glycerol and three fatty acids. Other lipids found in small amounts include mono- and diglycerides (esters of glycerol and one or two fatty acids, respectively), phospholipids such as lecithin, and sterols such as cholesterol. The last is synthesized only by animals.

Fatty acids are responsible for the varied characteristics of different triglycerides. Fatty acids are organic acids generally with an even number of carbon atoms. Triglycerides with short-chain fatty acids are usually liquid or soft at room temperature; the triglycerides become harder as chain lengths increase. Fatty acids may be saturated (having no double bonds), monosaturated

TABLE 10.1

High Intensity Sweeteners

Product	Sweeteners Index (Sugar=1)	FDA Status	Stability	Composition
Acesulfame-K	200	Approved	Stable	Derivative of (potassium) acetoacetic acid
Alitame	2000	Under review	Stable	Dipeptide of D-alanine and L-aspartic acid
Aspartame	180	Approved–must be labeled that it contains phenylalanine	Unstable when heated or when stored at pH extremes	Methylester of dipeptide of phenylalanine and aspartic acid
Cyclamate	30	Banned in U.S. in 1970	Stable	Sulfonated cyclohexylamine
Saccharin	300	Approved	Stable	Benzoic sulphimide
Sucralose	600	Under review	Stable	Chlorinated derivative of sucrose

(having one double bond), or polyunsaturated (having two or more double bonds). Saturated fatty acids are harder (more likely to be solid at room temperature) than unsaturated fatty acids, which tend to be liquid at room temperature.

Many commercially available fats are made from oils by hydrogenation. Hydrogenation is the process of adding hydrogen to unsaturated fatty acids under carefully controlled conditions to reduce the number of double bonds and increase saturation. Regardless of source and fatty acid composition, all fats and oils provide 9 kilocalories per gram.

ANIMAL FAT SHORTENINGS

These products are made from fat extracted from animal tissue by dry or wet rendering. The fats are then deodorized, and/or partially hydrogenated, and plasticized.

Lard—This is a plastic fat extracted from hogs. It is used in refried beans, pies, and pastries.

Tallow—This is fat extracted from sheep or cattle. It is more saturated than lard and has a more distinctive, meaty flavor. The flavor has made this a popular frying fat for french fries. Tallow currently makes up about 17 percent of the fat and oils used in shortening manufacture.

VEGETABLE OILS

Many of the edible oils used in the American home are of vegetable origin. Vegetable oils are pressed or squeezed from the seeds or fruits of the plant under heavy pressure (a method known as expelling), or the oils are dissolved

TABLE 10.2
Solubility of Sugars

Sugar	Temperature		Percentage of Sugar in Saturated Solution (%)	Amount Dissolved by 100 grams of Water (g)	Dissolved by 1 cup Water (236.6 ml) (calculated on the assumption that 1 cup sugar weighs 200g) (c)
	°F	°C			
Common Sugars					
Fructose	68	20	78.9	375.0	
Glucose	68	20	49.7	83.1	
Lactose	77	25	17.8	21.7	
Maltose	70	21	44.1	78.9	
Sucrose	32	0	64.2	179.2	2.1
	68	20	67.1	203.9	2.4
	104	40	70.4	238.1	2.8
	140	60	74.2	287.3	3.4
	176	80	78.4	362.1	4.3
	194	90	80.6	415.7	4.9
	212	100	83.0	487.2	5.8

out with an organic solvent that is later evaporated off (a method called solvent extraction). The raw oils are refined, bleached, and deodorized before being packaged. Oils to be used as salad oils are winterized by exposure to low temperatures for a period of time. These oils are then filtered to remove crystallized high-melting portions of the oil so that the remainder will stay clear at refrigerator temperatures.

Soybean, cottonseed, corn, peanut, safflower, and sunflower oils are widely used. All are bland in flavor and useful in a variety of food products. Safflower and sunflower oils, however, are poor frying oils due to their high unsaturation. Of increasing popularity is canola oil, which is extracted from a variety of rape seed and was approved for food use in 1985. Canola oil has a high content of monounsaturated fatty acids. This oil is 94 percent unsaturated. Only hydrogenated canola oil is suitable for frying.

Olive Oil

This oil is pressed from fully ripe black olives. "Extra virgin" and "virgin" olive oil are obtained from the first pressings. "Extra virgin" refers to oils with low acidity. Refined olive oil is derived from additional pressings of the fruit and is then filtered through layers of felt to remove impurities. Refined olive oil is not bleached or deodorized. Olive oil is popular because of its distinctive flavor and very high composition of monounsaturated fatty acids, a health benefit.

Tropical Oils

Palm oil, palm kernel oil and coconut oil are sometimes referred to as tropical oils. These oils are more saturated than other vegetable oils and are popular with food manufacturers because of their shelf stability.

Sesame Oil

Sesame oil ranges from a distinctively flavored straw- to

TABLE 10.3

Temperatures and Tests for Syrup and Candies

Product	Final Temperature of Syrup at Sea Level*		Test of Doneness	Description of Test
	°F	°C		
Jelly	220	104	—	Syrup runs off a cool metal spoon in drops that merge to form a sheet.
Syrup	230 to 234	110 to 112	Thread	Syrup spins in 2″ (5 cm) thread when dropped from fork or spoon.
Fondant Fudge Panocha	234 to 240	112 to 115	Soft ball	Syrup, when dropped into very cold water, forms a soft ball that flattens on removal from water.
Caramels	244 to 248	118 to 120	Firm ball	Syrup, when dropped into very cold water, forms a firm ball that does not flatten on removal from water.
Divinity Marshmallows Popcorn Balls	250 to 266	121 to 130	Hard ball	Syrup, when dropped into very cold water, forms a ball that is hard enough to hold its shape, yet plastic.
Butterscotch Taffies	270 to 290	132 to 143	Soft Crack	Syrup, when dropped into very cold water, separates into threads that are hard but not brittle.
Brittle Glacé	300 to 310	149 to 154	Hard Crack	Syrup, when dropped into very cold water, separated into threads that are hard and brittle.
Barley sugar	320	160	Clear liquid	The sugar liquefies.
Caramel	338	170	Brown liquid	The liquid becomes brown.

*For each increase of 500 feet in elevation, cook the syrup to a temperature 1°F lower than temperature called for at sea level. If readings are taken in Celsius, for each 900 feet of elevation cook the syrup to a temperature 1°C lower than called for at sea level.

amber-colored oil (found in Oriental food stores) to a more highly refined yellow product having a mild flavor (found in grocery stores). The oil is usually relatively expensive and is used primarily as a condiment for flavoring rather than for cooking. Sesame oil consists primarily of polyunsaturated fatty acids.

Marine Oils

These are oils extracted from herring, menhaden, whale, and sardines. Only menhaden oil is currently approved for use in the U.S. These oils are the subject of current research since they are high in omega-3 fatty acids, unsaturated fatty acids linked with a decreased risk of heart disease.

HYDROGENATED ALL-VEGETABLE SHORTENINGS

Essentially these are solidified vegetable oils. Soybean oil is used primarily, but varying amounts of other oils such as corn oil, peanut oil and cottonseed oil are sometimes used.

The vegetable oils are refined to remove free fatty acids, and then bleached with absorbent materials to remove coloring materials. The purified oils are then hydrogenated and deodorized to produce a bland-flavored fat. Mono- and diglycerides are added to enhance their emulsifying capability.

SPREADS

Butter

See page 64 (In Dairy Section).

Margarine

Margarines are made from refined vegetable oils or a combination of animal fats and vegetable oils emulsified with cultured milk, sweet milk, nonfat dry milk solids, water, or a mixture of these. The emulsion is then cooled and kneaded by machine to produce the desired consistency. Today almost all margarines are enriched with carotenes to enhance color. Salt and butter flavoring may be added. The law requires that margarine contain 80 percent fat unless the product is intended as a diet substitute, in which case the package must be labeled "imitation" or "diet." Ingredient labels must also state the type of fat or fats used.

Regular margarine varies in hardness due to the type of oil or fat used and the amount of hydrogenation. Spreads and liquid margarines generally are higher in polyunsaturated fatty acids than regular stick margarine. Both regular and soft margarines provide 100 Kcal/Tbsp. (15 ml) and contain no cholesterol if produced from vegetable oils. Whipped margarines have added air and provide about 60 kcal/Tbsp. (15 ml). Diet margarines contain 40–45 percent fat and added water; they provide 50–80 percent fewer calories than regular margarine. They should not be substituted in regular recipes for the 80 percent fat products.

Blends

Blends are mixtures of margarine and butter. The products offer a butter-like taste with one-third to one-sixth less cholesterol than butter.

FAT SUBSTITUTIONS

Oils may be substituted for solid shortening in batters and doughs by reducing the amount of oil 1½ tablespoons (22.5 ml) per cup (236.6 ml). However, a product made with oil will be different from that made with shortening. Pastries will be more crumbly and mealy in texture because the oil is spread homogeneously throughout the flour mixture. Cakes prepared by the creaming or traditional method are lower in volume and more coarse when oil is used as a replacement for shortening. Less air is incorporated during the creaming stage of the shortening and sugars, reducing the number of air cells that expand and divide during the baking process. To use oil in a cake, the mixing can be modified to a muffin method whereby the oil is added with other liquids; an egg white foam folded into the batter at the end of mixing helps to increase the volume.

FAT SUBSTITUTES

Modified foods that may be labeled light, reduced calorie, or reduced fat are available in the market. The foods may be of a fat-reduced formula or contain a fat substitute that is either not metabolized or has an energy value of 4 kilocalories (17kJ) or less per gram compared to the 9 kilocalories (38 kJ) per gram for fats.

Protein-based fat substitutes such as Simplesse® are made

from egg white and milk protein that have been heated and formed into tiny spheres, by a process called microparticulation. This gives the creamy mouthfeel of fat to frozen products, cream cheese, cheese spreads, or yogurt.

Carbohydrate-based fat substitutes are combinations of gums and other carbohydrates such as polydextrose that act as a bulking agent. N-Oil® is made from tapioca dextrin and maltodextrins made from hydrolyzed cornstarch. These products contribute to viscosity and impart the mouthfeel of fat. Both protein and carbohydrate substitutes have water blended in ratios that result in a final product that may range from two to four kilocalories (eight to 17 kJ) per gram.

Synthetic-based fat substitutes include triglyceride modification (sucrose polyester), polycarboxylic acid esters (trialkoxy-tricarballylate or TATCA by Best Foods) sterically hindered esters and polyglycerines. Triglyceride modifications are composed of modified molecules of sugars linked with long chain fatty acids. The synthetic substitutes have the look and feel of fat but the large molecules are not metabolized and, therefore, have no caloric value. Currently, such a product, Olestra®, has been approved by the Food and Drug Administration for use in replacing fat in salted snack foods. Olean is made from sucrose polyester and is used in various snack chips.

STORAGE AND USE

- Store lards and home-rendered fats such as poultry fat in the refrigerator.
- Refrigerate vegetable shortenings intended for storage of several months or more. These fats, however, will keep well at room temperature for shorter periods of time.
- Keep oils capped tightly and store at room temperature. Some oils may become thick and cloudy when stored in the refrigerator due to partial solidification Store in dark containers or in a dark storage area.

FRYING WITH FATS AND OILS

Fats and oils are used as a cooking medium for pan-frying or deep fat frying. Oils may be preferred since they

generally are more stable and have a higher smoke point than solid shortenings. Some shortenings contain mono- and diglycerides that are desirable for use in cakes, but contribute a lower smoke point. The smoke point is the lowest temperature at which a continuous stream of thin, visible smoke is observed, along with a strong odor and eye irritation. The smoke is symptomatic of a rapid breakdown of the fat, resulting in off flavor to the food cooked in the fat. The smoke point of the oil/fat must be above the temperature used in food preparation. The smoke point is lowered greatly by the presence of free fatty acids.

The formation of free fatty acids and other products from fat deterioration are accelerated with exposure to light, food particles in the fat, water from the food cooked in the fat, and a wide container with a large surface area. To maintain high quality of fat or oil to be reused for deep fat frying, the oil should be strained and stored in a closed container in cool darkness. Approximately 1/5 of the fat should be replaced with new fat or oil for each use.

SMOKE POINTS OF FATS AND OILS

Lard	361°– 401° F (183°– 205°C)
Vegetable Oil	441°– 450°F (227°– 232°C)
Vegetable shortenings with mono- and diglycerides	356°– 370°F (180°– 188°C)
Vegetable and animal shortenings with mono- and diglycerides	351°– 363°F (177°– 184°C)
Vegetable and animal shortenings without mono- and diglycerides	448°F (231°C)

Fat absorption is reduced by maintaining a high enough oil cooking temperature to allow the food to cook quickly (but not brown so rapidly that the interior is undercooked). Other factors that increase fat absorption are increased amounts of fat and sugar in the batter of the product, a lower protein content of the mixture used, and greater surface area of the product.

WAYS TO MEASURE FATS AND OILS

Butter or margarine purchased in bar form need not be measured with measuring equipment. Simply keep in mind that a $1/4$ pound bar equals $1/2$ cup (125 ml) or 8 tablespoons (120 ml). Two bars equal 1 cup (250 ml).

For other fats not in bar form use standard measuring cups or spoons. Press fat firmly into the utensil until it is full. Level with the straight edge of a spatula or knife.

Water displacement method may be used if water that clings to the fat will not affect the product. Pour cold water into a liquid measuring cup up to the measure that will equal 1 cup (250 ml) when the desired amount of fat is added. For example, if $1/4$ cup (60 ml) is needed, pour $3/4$ cup (175 ml) water into the measurer. Add enough fat to the water to make the water level rise to the mark for 1 cup (250 ml), being sure that the fat is entirely covered with water. Drain off the water.

For oils or melted fats, use a standard glass measuring cup or spoons and pour the oil or melted fat to the desired mark.

LEAVENING AGENTS

A leavening agent is a gas incorporated or formed in a batter or dough to make it rise, increase in volume or bulk, and become light and porous during preparation and subsequent heating. The amount of leavening gas in a mixture, the rate at which it is formed, and the flavors it imparts are important in establishing texture and other characteristics of leavened products. There are three principle leavening gasses:

Air

Air contributes to some of the volume of leavened products. Air is beaten or folded into mixtures or introduced into ingredients by beating, creaming and sifting. It leavens by expansion during heating.

Water Vapor

Water vapor or steam is formed in any batter or dough as it is heated. It is the principle leavening agent in products having a high proportion of liquid to flour, such as popovers and cream puffs. To form steam before the product coagulates or sets, a high initial temperature must be reached.

Carbon Dioxide

Carbon dioxide is a leavening agent produced in a batter or dough by chemical or biological reactions. Currently it is produced in breads and cakes from baking soda (sodium bicarbonate) in combination with acid ingredient(s), baking powder, yeast or other microorganisms metabolizing sugar.

FROM SODIUM BICARBONATE AND ACIDS

Baking soda plus an acid ingredient form carbon dioxide gas, salts and water. Common sources of acids are sour or acidified milk, buttermilk, sour cream, citrus fruit juice, vinegar, applesauce, molasses, and brown sugar. The acidity of sour milk varies with its age and degree of sourness; the acidity of molasses also varies. The acidity of corn syrup, honey and chocolate is too low for them to be used as the only source of acid. Then baking powder is also added. When sour milk or molasses is used, the baking soda should be mixed with the dry ingredients, since the reaction between baking soda and acid is immediate in a liquid medium. For proportions of baking soda and acid to use in place of baking powder see "Substitution of Ingredients," pages 31–32.

FROM BAKING POWDERS

Baking powders are mixtures of dry acid or acid salts and an alkali (sodium bicarbonate) separated by an inert filler such as cornstarch or powdered calcium carbonate added to standardize and help stabilize the mixtures. According to federal standards, baking powders must liberate at least 12 percent of available carbon dioxide. Baking powders are classified according to type of acid components. Those commonly found in the supermarket for home use are SAS and low-sodium.

- Tartrate powders in which the acid ingredients are potassium acid tartrate (cream of tartar) and tartaric acid. These are quick acting or single acting baking powders that form gas bubbles as soon as the batter is mixed.

- Phosphate powders, in which the acid ingredient is either calcium acid phosphate or sodium acid pyrophosphate or a combination of these. Phosphate powders may be double acting, meaning some reaction occurs at room temperature and the remaining reaction occurs with heat.

- Sodium-free powders or low sodium powder, may be double acting in which potassium bicarbonate replaces the sodium bicarbonate. Use 1.5 teaspoonful (7 ml) for each teaspoonful of baking powder in the recipe.

- SAS-phosphate powders, in which the acid ingredients are sodium aluminum sulfate and calcium acid phosphate. This type is often referred to as combination or double-acting powder and may be so considered from the point of view that the acid phosphate reacts with the baking soda while the mixture is cold, whereas the sodium aluminum sulfate produces carbon dioxide when the batter is heated. This type is always available in supermarkets and used in home baking.

Tartrate powders release essentially all of their carbon dioxide while the mixture is cold and SAS-phosphate powders release most of their carbon dioxide when heated; phosphate powders are intermediate. Different amounts of carbon dioxide are lost during mixing, depending on the type of baking powder used.

FROM BIOLOGICAL AGENTS

Carbon dioxide is formed by the action of yeast or certain bacteria with sugar. These reactions require a fermentation or proofing period prior to baking in order to produce the leavening gas.

Yeast—a microscopic, unicellular plant—Under suitable conditions of temperature, nutrients, and moisture, it produces carbon dioxide from simple sugars formed from starch and/or granulated sugar. Yeast is marketed in three forms, which are interchangeable:

Compressed yeast is a moist mixture of yeast and starch.

The yeast is in an active state. Presence of approximately 70 percent moisture makes the product perishable; therefore, it should be refrigerated.

Active dry yeast is similar to compressed yeast except the yeast-and-filler mixture has been dried and is then packaged in granular form. It should be rehydrated before mixing with the flour in making a dough.

Quick rising yeast (rapid rising or instant) is a more active strain of yeast packaged in a dried form. It does not need to be rehydrated before mixing and is, therefore, more suitable for use in a home bread machine. For a one pound loaf, 2.5 teaspoons may be better than a whole package. Up to 50 percent less rising time is needed with this yeast than with active dry yeast.

Bacteria of certain species, under suitable conditions of temperature and moisture, grow rapidly and produce gases from sugar. Salt-rising and sourdough breads are made form dough leavened in this manner. These doughs will have a distinctive taste.

Acidic (sour) doughs contain *Saccharomyces exigus* and *Saccharomyces inusitatus*. The high acidity of the dough results in the distinctive flavor of sour dough. San Francisco sourdough bread uses in addition *Lactobacillus sanfrancisco* that produces lactic acid. Other breads made from sourdough include rye breads and pumpernickel.

STORAGE AND USE

- Store baking powder, baking soda, and cream of tartar tightly covered in a dry place.

- Check the label for the expiration date, and plan to use it before that date.

- Before measuring dry leavening agents, stir the product to break up any lumps. For best results, use standard measuring spoons. Be sure the spoon is dry when the product is measured.

Miscellaneous Foods

DEFINITIONS

Bouillon

Bouillon is a clarified liquid prepared by simmering meat or poultry.

Bouillon cube or powder is a dehydrated cube or powder containing hydrolyzed vegetable protein, salt, fat, dextrin, sugar, meat extract, flavorings and coloring.

Bread Crumbs

Dry breadcrumbs are those that can be rolled fine. They are used for stuffing, for buttered crumbs, and for coating foods for frying. Packaged breadcrumbs are of this type.

Soft breadcrumbs are those prepared by crumbling 2- to 4-day-old bread. These are used for bread puddings, fondues, timbales, stuffing, and buttered crumbs.

Croutons are dried bread cubes sold commercially. They may have a variety of herb and spice mix coatings.

Catsup

Catsup, catchup, or ketchup is prepared from concentrated tomato pulp and liquid, seasoned with onions and/or garlic, salt, vinegar, spices, and/or flavorings, and sweetened with sugar, dextrose, or corn syrup.

Chili Sauce

Chili sauce is similar to catsup but contains pieces of the whole peeled tomato with seeds and more sugar and onion than does catsup.

Chocolate

Chocolate is the product resulting from the grinding of cocoa nibs (cocoa, or cacao, beans that have been roasted and shelled).

Sweet Chocolate (sweet chocolate coating) is chocolate mixed with sugar and may also contain added cocoa butter and flavorings. It is used for dipping confections.

Semisweet chocolate pieces or squares are formed from slightly sweetened chocolate.

Unsweetened chocolate is the original baking or cooking chocolate with no sweeteners or flavorings added.

White chocolate is milk chocolate that contains mild flavored cocoa butter (the fat of the cocoa bean), but no additional cocoa solids. White chocolate keeps for a shorter time than the more familiar chocolates.

Cocoa

Cocoa is powdered chocolate from which a portion of the cocoa butter has been removed.

Breakfast cocoa is a high-fat cocoa that must contain at least 22 percent fat.

Cocoa has a medium fat content (from 10 to 21 percent cocoa fat).

Dutch process cocoa can be either "breakfast cocoa" or "cocoa" which is processed with one or more alkaline materials, as permitted under government regulations.

Instant cocoa is a mixture of cocoa, sugar, and an emulsifier. It can be prepared as a beverage by adding hot liquid.

Coconut

Flaked or grated coconut is coconut meat cut into uniform shreds or flakes.

Coffee

Coffee is prepared by roasting green coffee beans, and then blending, and usually grinding. Flavor of the brewed beverage depends on the degree of roasting. In some parts of the country—notably Louisiana—coffee blended with chicory is favored.

Instant coffee is prepared by freeze-drying or by various extraction, evaporation, and drying processes.

Decaffeinated coffee is prepared either by steaming and soaking green coffee with a chlorinated organic solvent or by a water-soaking process.

Fruit Pectin

Fruit pectin is part of the water-soluble fiber found in fruit. In the right proportion with sugar and acid, pectin forms a gel. Commercially, pectin is refined from citrus or apple pectin and marketed as liquid or powdered pectin.

Gelatin

The term gelatin usually means the granulated, unflavored, unacidulated product. Gelatin is obtained by hydrolysis of collagen in bones and good-grade skin stock. In processing, gelatin may be alkaline- or acid-extracted.

Fruit-flavored gelatin is a mixture of plain gelatin, sugar or sugar substitute, fruit acids, flavors, and coloring. It is sold in packages standardized to gel 1 pint or 1 quart of liquid (500 – 1000 ml). For industrial use, this product is packaged in one-pound (454 g) or larger containers.

Infant Foods

A wide variety of strained and junior or chopped foods is available. Infant foods are also labeled as beginner, second food, third food and graduates. Food is preserved in small containers, usually glass, of cooked cereals, strained or chopped fruits, strained or chopped vegetables, strained meats, and combinations or mixtures of these foods. In addition, frozen dinners (usually of meat, vegetable, and pasta) are available for infants, toddlers and juniors.

Mayonnaise and Salad Dressings

Mayonnaise is a permanent emulsion of oil droplets in water, stabilized with egg yolk. It is prepared from vegetable oil, vinegar or lime or lemon juice, eggs or egg yolks, and spices. Commercial mayonnaise must contain a minimum of 65 percent vegetable oil.

Salad dressings have substantially the same ingredients as those in mayonnaise, but a portion of the egg is replaced with a cooked starch paste, and the amount of oil is less than in mayonnaise. Salad dressing is 30 percent vegetable oil.

Some French dressings are temporary emulsions. Other commercial French dressings are emulsified with small amounts of vegetable gums or pectins. These dressings also contain tomato paste or purée and a minimum of 35 percent vegetable oil.

Low-calorie or fat-free dressings may have a fruit or vegetable base, and contain water, sugar, vinegar, modified starches, spices, vegetable gums, and emulsifiers, with very little or no oil. They may also be sweetened artificially.

Mustard

A pungent condiment consisting of black and/or yellow mustard seeds pulverized and made into a paste with water and/or vinegar. The paste may then be mixed with spices, sugar, and/or salt. Ground mustard seeds are available as a dry powdered spice.

Nuts

Nuts are dry fruits that generally consist of a single kernel inside a woody shell. True nuts include filberts, hazelnuts, almonds, pecans, Brazil nuts and black walnuts. Cashews and pistachios are nuts (seeds) that grow on small trees. Pine nuts were commonly consumed by Native Americans of the Southwest but are now available in stores. Peanuts are the pods of a vine of the pea family and are classified as a legume.

Nuts are available either in the shell or shelled. Shelled nuts may be chopped, ground, blanched, halved, slivered, plain, toasted, and/or salted.

Peanut butter is a spread prepared from finely ground nuts which may be blanched or unblanched. Commercially prepared peanut butter may be hydrogenated and contain salt and/or a sweetener.

Olives and Olive Oil

The edible fruit of the olive tree is available in cans or jars as ripe olives, green fermented, or salt-cured olives, as well as an oil.

Both green and ripe olives are treated to remove the characteristic bitterness of the nut. Ripe olives are packed in salt with or without spices and are available pitted, unpitted, whole, sliced, or chopped.

Green olives are fermented, and packed in brine, either whole, pitted, or pitted and stuffed with pimiento, almonds, capers, onions, or celery.

Dried or salt-cured olives are also known as Green or Italian olives.

The U.S. grades for ripe or green olives are Grade A (Fancy), Grade B (Choice), Grade C (Standard), and Substandard.

Green olives are available in the following sizes. Ripe olives have a similar size range:

No. 1 (small)	128 to 140 per lb
No. 2 (medium)	106 to 127 per lb
No. 3 (large)	91 to 105 per lb
No. 4 (extra large)	76 to 90 per lb
No. 5 (mammoth)	65 to 75 per lb
No. 6 (giant)	53 to 64 per lb
No. 7 (jumbo)	46 to 52 per lb
No. 8 (colossal)	33 to 45 per lb
No. 9 (super colossal)	32 maximum

Pickles

Pickles are cucumbers, other vegetables, or fruits, prepared by fermentation or in vinegar, usually with salt, sugar, and spices added. There are three general groups: (1) fermented (salt and dill pickles); (2) unfermented (fresh pasteurized); and (3) sweet, sour, and mixed pickles and relishes of various mixtures.

Pudding

Instant pudding comes in a variety of flavors. It generally contains a modified starch, milk solids, surfactants and a variety of flavorings and colorings.

Salsa

Salsa is a combination of chopped onion, cilantro, and tomato which may contain other ingredients and spices commonly served to enhance the flavor of Mexican cuisine.

Salt

Salt is sodium chloride unless otherwise identified as, for example, potassium chloride. Salt (sodium chloride) is used to season and preserve foods. Sometimes spices and/or herbs or other seasonings are added as in onion, garlic, or celery salt. Salt may also be iodized.

Salt is made "free flowing" by the addition of a substance (sodium silico aluminate) that prevents absorption of moisture from the air.

Pickling salt differs from common salt in that pickling salt does not contain additives that would cloud the pickle liquid. Pickling salt may be granulated or flaked. One cup (250 ml) of the granulated salt is equal to 1$\frac{1}{2}$ cups (375 ml) of flake salt.

Rock salt is a granular salt used in salt mills and in ice cream making.

Salt substitutes contain calcium, potassium, or ammonium in place of sodium.

Soups

Soups may be clear or thick and may be served hot or cold. Bisque is a rich cream soup sometimes containing shellfish. Bouillabaisse is a fish soup or stew. Bouillon or broth is a thin soup prepared by simmering meat, fish, or vegetables in water to extract their flavor. Consommé is a clarified, double strength brown broth. Chowder is a heavy thick soup prepared from meat, poultry, fish, and/or vegetables. Fruit soups are served as a dessert. Gazpacho is a cold soup of fresh vegetables and herbs, including tomatoes. Vichyssoise is a cream stock of puréed potatoes, leeks and chicken stock served cold.

Spices and Herbs

Most spices are grown in tropical climates and herbs in temperate climates. Spices and herbs are used to season food and include various plant parts that have aromatic odors and pungent flavors. They include:

Aril (a lacy layer of the nutmeg seed)	Mace
Bark	Cassia, cinnamon
Berry	Allspice, juniper, peppercorn
Bud	Capers, cloves
Flower Stigma	Saffron
Fruit	Cayenne pepper, paprika
Kernels or seeds	Anise, caraway, cardamom, celery, coriander, cumin, dill, fennel, fenugreek, mustard, nutmeg, poppy, sesame
Leaves and stems	Basil, bay, chervil, celery, chives, dill weed, marjoram, mint, parsley, oregano, rosemary, sage, savory, tarragon, thyme
Root	Ginger, horseradish, turmeric

Tapioca

Tapioca is made from flour obtained from the cassava root and is marketed in two forms:

Pearl tapioca consists of small pellets that thicken and become translucent in cooking. Pearl tapioca is made by mixing the tapioca flour with water and cooking it on heated metal surfaces just enough to form a shell on the pellets.

Quick cooking tapioca consists of very fine pellets. It is made by grinding a cooked dough prepared from tapioca flour or by crushing pearl or native flake tapioca.

Flavored tapioca mixes are blends of quick-cooking tapioca, cornstarch, sugar and flavorings.

Tea

Tea is prepared from the leaves of an evergreen tree or shrub (*Thea sinensis*). The treatment of the leaves after they are picked produces three types of teas. Black tea is prepared by allowing dried and rolled tea leaves to ferment before they are fired. Green tea is steamed, rolled, dried, and fired without fermentation. Oolong tea has partially fermented leaves. For special teas, including herbal, the leaves are mixed with jasmine, gardenia, mint, orange, or other spices.

Comfrey, as a tea, is a health hazard when taken internally because of toxic pyrrolizidine alkaloids shown to be carcinogenic in rats.

Chamomile and *yarrow* have long been used as teas for digestive disorders. Those allergic to ragweed, asters, or chrysanthemums should drink these with caution.

Sassafras has been used as a tea; it contains safrole, a volatile oil, and a carcinogenic agent in rats.

Textured Vegetable Proteins

Textured vegetable protein products are made from edible protein sources. Extruded products are colored and flavored to resemble a food such as ground beef and are most often used as extenders. Spun soy products are also colored and flavored and shaped to resemble products such as ham, chicken cubes, beef or bacon pieces. Also available are frozen, canned, and dehydrated spun soy products.

Vinegars

The acidity of vinegars ranges between four and six percent (40 to 60 grain).

Vinegar or cider vinegar is the product made by fermenting apple juice.

Malt vinegar is the product made by the alcoholic and subsequent fermentations of an infusion of barley malt or cereals whose starch has been converted by malt.

Wine vinegar is the product made by the alcoholic and subsequent acetous fermentations of the juice of grapes.

Spirit, distilled, or grain vinegar is made by the acetous fermentation of dilute distilled alcohol.

STORAGE AND USE

- Store nuts tightly covered in a cool, dry, dark place or in the freezer. Exposure to air, light, warmth, and moisture can cause rancidity.

- Store spices in tightly closed containers in a cool, dry place. Refrigeration or freezing extends their shelflife.

- Keep mayonnaise and salad dressings made with eggs in the refrigerator once the jar has been opened.

- Keep gelatin in unopened package until ready for use, and store in a dry place.

- To substitute fresh herbs for dried, use 2 teaspoons (10 ml) minced fresh herbs for each 1/4 teaspoon (1 ml) of the dried product.

- Melt chocolate for cooking purposes in small container over hot water unless recipe states otherwise. Chocolate scorches easily when exposed to direct heat. It can be melted in the microwave oven.

- Follow storage directions when provided on package labels.

Food Buying Guides

Buying Guide for Fats and Oils

Food Item and Form	Market Unit		Approximate Volume per Market Price		Approximate Weight per Cup	
Butter (See Dairy Products)						
Oils: corn, cottonseed, olive, peanut, and safflower	1 qt	946 ml	4 c	946 ml	218 g	7.7 oz
Margarine	1 lb	454 g	2 c	473 ml	227 g	8.0 oz
Whipped	1 lb	454 g	3 c	710 ml	150 g	5.3 oz
Soft tub	8 oz	227 g	1 c	237 ml	227 g	8.0 oz
Imitation (40% fat)	1 lb	454 g	2 c	473 ml	232 g	8.2 oz
Hydrogenated fat	1 lb	454 g	2½ c	551 ml	204 g	7.2 oz
Lard	1 lb	454 g	2⅓ c	551 ml	204 g	7.2 oz
Beef tallow	1 lb	454 g	3¾ c	887 ml	204 g	7.2 oz
Mutton tallow					204 g	7.2 oz

Buying Guide for Poultry

Food Item and Form	Market Unit		Approximate Number of Servings per Market Unit*	Approximate Weight per Cup	
Chicken					
Ready-to-cook					
Broiler-fryer	1 lb	454 g	2		
Roaster	1 lb	454 g	2		
Rock Cornish hen	1 lb	454 g	1		
Capon	1 lb	454 g	2		
Stewing					
Cooked, boned	1 lb	454 g	2		
diced			1½	139 g	4.9 oz
Canned, boned	5 oz	140 g	1½ to 2		
Duck					
Ready-to-cook	1 lb	454 g	1		

Buying Guide for Poultry (continued)

Food Item and Form	Market Unit		Approximate Number of Servings per Market Unit*	Approximate Weight per Cup	
Goose					
Ready-to-cook	1 lb	454 g	1$^{1}/_{2}$ to 2		
Turkey					
Ready-to-cook					
bone in	1 lb	454 g	2		
boneless roast	1 lb	454 g	3		
ground	1 lb	454 g	4		
cooked, boned,					
diced,	5 oz	140 g	1 to 2	139 g	4.9 oz
canned, boned	5 oz	140 g	1 to 2		

*Amounts are based on three ounces of cooked poultry meat without bone per serving.

Buying Guide for Eggs

Food Item and Form	Market Unit		Approximate Volume or Number per Market Unit		Approximate Weight per Cup	
Eggs, Whole						
Fresh	1 doz		12 eggs		244 g	8.6 oz
Jumbo, 63 g	1 doz		3 c	710 ml	244 g	8.6 oz
Extra large, 56 g	1 doz		2$^{3}/_{4}$ c	651 ml	244 g	8.6 oz
Large, 50 g	1 doz		2$^{1}/_{4}$ c	532 ml	244 g	8.6 oz
Medium, 44 g	1 doz		2 c	473 ml	244 g	8.6 oz
Small, 38 g	1 doz		1$^{7}/_{8}$ c	444 ml	244 g	8.6 oz
Frozen	1 lb	454 g	1$^{7}/_{8}$ c	444 ml	244 g	8.6 oz
Dried, sifted	1 lb	454 g	5$^{1}/_{3}$ c	1.3 l	82 g	2.9 oz
Whites						
Fresh	1 doz		12 whites		244 g	8.6 oz
Jumbo, 42 g	1 doz		1$^{7}/_{8}$ c	473 ml	244 g	8.6 oz
Extra large, 37.4 g	1 doz		1$^{2}/_{3}$ c	444 ml	244 g	8.6 oz
Large, 33.4 g	1 doz		1$^{1}/_{2}$ c	393 ml	244 g	8.6 oz
Medium, 30.4 g	1 doz		1$^{1}/_{4}$ c	355 ml	244 g	8.6 oz
Small, 25.4 g	1 doz		1$^{7}/_{8}$ c	296 ml	244 g	8.6 oz
Frozen	1 lb	454 g	1$^{7}/_{8}$ c	444 ml	244 g	8.6 oz
Dried, sifted	1 lb	454 g	4$^{1}/_{4}$ c	1.0 l	108 g	3.8 oz

Buying Guide for Eggs (continued)

Food Item and Form	Market Unit		Approximate Volume or Number per Market Unit		Approximate Weight per Cup	
Yolks						
Fresh	1 doz		12 yolks		244 g	8.6 oz
Jumbo, 21 g	1 doz		1 c	237 ml	244 g	8.6 oz
Extra large, 18.6 g	1 doz		7/8 c	207 ml	244 g	8.6 oz
Large, 16.6 g	1 doz		5/6 c	197 ml	244 g	8.6 oz
Medium, 14.6 g	1 doz		2/3 c	156 ml	244 g	8.6 oz
Small, 12.6 g	1 doz		5/8 c	148 ml	244 g	8.6 oz
Frozen	1 lb	454 g	1 7/8 c	444 ml	244 g	8.6 oz
Dried, sifted	1 lb	454 g	6 3/4 c	1.6 l	65 g	2.3 oz
Substitutes	16 oz	454 g	2 c	473 ml	227 g	8 oz

Buying Guide for Fish and Shellfish

Food Item and Form	Market Unit		Approximate Servings per Market Unit*	Approximate Weight per Cup	
Fish, Fresh or Frozen					
Whole	1 lb	454 g	1 1/2		
Chunks	1 lb	454 g	3		
Dressed	1 lb	454 g	2 1/3		
Fillets	1 lb	454 g	3 1/3		
Steaks	1 lb	454 g	3		
Cakes, frozen	1 lb	454 g	5 1/3		
frozen portions					
unbreaded	1 lb	454 g	4		
breaded, fried or raw	1 lb	454 g	5 1/3		
Sticks, frozen	1 lb	454 g	5 1/3		
Fish, canned					
Gefilte fish	1 lb	454 g	3	162 g	5.7 oz
Mackerel	15 oz	424 g	4 1/4	181 g	6.4 oz
Sardines	12 oz	340 g	3 3/4	159 g	5.6 oz
Salmon	1 lb	454 g	4 1/4	167 g	5.9 oz
Tuna	7 oz	198 g	2	170 g	6.0 oz

Buying Guide for Fish and Shellfish (continued)

Food Item and Form	Market Unit		Approximate Servings per Market Unit*	Approximate Weight per Cup	
Fish, cured					
Lox	1 lb	454 g	$5^1/_3$		
Salt fish	1 lb	454 g	$5^1/_3$		
Smoked fish	1 lb	454 g	$3^1/_2$		
Shellfish					
Clams					
in shell (hard)	1 doz		2		
in shell (soft)	1 doz		1		
shucked	1 lb	454 g	$2^1/_2$		
frozen, breaded, raw	1 lb	454 g	$4^1/_2$		
canned, minced	7.5 oz	213 g	$2^1/_2$	159 g	5.6 oz
Crabs					
in shell (Blue)	1 lb	454 g	$^3/_4$		
in shell (Dungeness)	1 lb	454 g	$1^1/_4$		
crab meat	1 lb	454 g	5	162 g	5.7 oz
crab cakes, frozen	1 lb	454 g	5		
crab legs and sections, frozen	1 lb	454 g	$2^1/_2$		
deviled, frozen	1 lb	454 g	$5^1/_3$		
crab meat, canned	6.5 oz	184 g	$1^3/_4$		
Lobsters					
in shell	1 lb	454 g	$1^1/_4$		
meat	1 lb	454 g	$4^3/_4$	153 g	5.4 oz
spiny tails, frozen	1 lb	454 g	$2^2/_3$		
Oysters					
in shell	1 doz		2		
shucked	1 lb	454 g	2	235 g	8.3 oz
breaded, frozen	1 lb	454 g	$4^2/_3$		
canned	5 oz	142 g	$1^2/_3$	156 g	5.5 oz
Scallops					
shucked	1 lb	454 g	$3^1/_3$		
breaded, frozen	1 lb	454 g	4		
Shrimp					
in shell	1 lb	454 g	$2^2/_3$		
raw, peeled	1 lb	454 g	$3^1/_3$		
cooked, peeled	1 lb	454 g	$5^1/_3$		
breaded, frozen	1 lb	454 g	$4^1/_2$		
canned	13.25 oz	376 g	$4^1/_3$	130 g	4.6 oz

*One serving equals three ounces (85 g) of cooked boneless fish or shellfish.

Buying Guide for Meats

Meat Form	Market Unit		Volume/Number Servings per Market Unit*
Fresh or Frozen			
Boneless or Ground Meat	1 lb	454 g	4 servings
Meat with minimum amount of bone			
(steaks, roasts, chops, etc.)	1 lb	454 g	2 to 3 servings
Meat with large amount of bone			
(shoulder cuts, short ribs, neck, etc.)	1 lb	454 g	1 to 2 servings
Cured and/or Smoked			
Ham, boneless	1 lb	454 g	4 to 5 servings
Bacon	1 lb	454 g	24 slices
Frankfurters	1 lb	454 g	8 to 10 frankfurters

*Three ounces (85 g) of cooked trimmed meat is the usual amount for one serving.

Buying Guide for Dairy Products

Food Item and Form	Market Unit		Approximate Volume per Market Price		Approximate Weight per Cup	
Butter	1 lb	454 g	2 c	473 ml	224 g	7.9 oz
Whipped	1 lb	454 g	3 c	710 ml	152 g	5.2 oz
Cheese						
Cheddar (natural						
or processed)	1 lb	454 g				
grated or						
chopped			4 c	946 ml	113 g	4.0 oz
Cheddar or Swiss						
sliced	1 lb	454 g	8 slices			
Cottage	12 oz	340 g	1½ c	355 ml	236 g	8.3 oz
Cream	8 oz	227 g	1 c	237 ml	230 g	8.1 oz
Spread	5 oz	142 g	½ c	118 ml		
Parmesan, grated	3 oz	85 g	1 c	237 ml	92 g	3.3 oz

Buying Guide for Dairy Products (continued)

Food Item and Form	Market Unit		Approximate Volume per Market Price		Approximate Weight per Cup	
Cream						
Light (table)	¹/₂ pt	237 ml	1 c	237 ml	240 g	8.5 oz
Heavy						
(whipping)			1 c	237 ml	236 g	8.3 oz
whipped	¹/₂ pt	237 ml	2 c	473 ml		
Sour	¹/₂ pt	237 ml	1 c	237 ml	241 g	8.5 oz
Half and Half						
(cream + milk)						
sweet	1 pt	473 ml	2 c	473 ml	242 g	8.5 oz
sour	¹/₂ pt	237 ml	1 c	237 ml	242 g	8.5 oz
Milk						
Whole or skim	1 qt	946 ml	4 c	946 ml	242 g	8.5 oz
Buttermilk	1 qt	946 ml	4 c	946 ml	242 g	8.5 oz
Sweetened						
condensed	15 oz	425 g	1¹/₃ c	315 ml	306 g	10.8 oz
Evaporated						
whole or skim	14.5 oz	411 g	1²/₃ c	393 ml	252 g	8.9 oz
reconstituted			3¹/₃ c	788 ml		
Dry, nonfat	1 lb	454 g	3²/₃ c	866 ml	131 g	4.6 oz
instant	9⁵/₈ lb	4.4 kg	4 c	946 ml	75 g	2.6 oz
reconstituted			14 c	3.3 l	242 g	8.5 oz
Milk desserts						
Ice cream	1 qt	946 ml	4 c	946 ml	142 g	5.0 oz
brick, sliced	1 qt	946 ml	8 slices			
Ice milk	1 qt	946 ml	4 c	946 ml	187 g	6.6 oz
Sherbet	1 qt	946 ml	4 c	946 ml	193 g	6.8 oz
Yogurt	¹/₂ pt	237 ml	1 c	237 ml	246 g	8.7 oz

Buying Guide for Cereals and Flours

Food Item and Form	Market Unit		Approximate Volume or Pieces per Market Unit		Approximate Weight per Cup	
Cereals/Grains						
Amaranth	1 lb	454 g	2½ c	592 ml	179 g	6.3 oz
Bulgur	1 lb	454 g	2¾ c	651 ml	162 g	5.7 oz
cooked			8 c	1.9 l	230 g	8.1 oz
Cornmeal						
white	1 lb	454 g	3½ c	828 ml	130 g	4.6 oz
yellow	1 lb	454 g	3 c	710 ml	153 g	5.4 oz
cooked			16⅔ c	3.9 l	238 g	8.4 oz
Farina	1 lb	454 g	3 c	710 ml		
cooked			16⅔ c	3.9 l	238 g	8.4 oz
Hominy, whole	1 lb	454 g	2½ c	592 ml	181 g	6.4 oz
cooked			16⅔ c	3.9 l		
Grits	1 lb	454 g	3 c	710 ml	153 g	5.4 oz
cooked			10 c	2.4 l	235 g	8.3 oz
Oats, rolled	1 lb	454 g	6¼ c	1.5 l	71 g	2.5 oz
cooked			8 c	1.9 l	241 g	8.5 oz
Quinoa, whole	14 oz	397 g	2½ c	592 ml	159 g	5.6 oz
Ready-to-eat						
flaked					31 g	1.1 oz
granulated					88 g	3.1 oz
puffed					23 g	0.8 oz
shredded					37 g	1.3 oz
Rice, brown	1 lb	454 g	2½ c	592 ml	184 g	6.5 oz
Rice, white polished						
long grain	1 lb	454 g	2½ c	592 ml	181 g	6.4 oz
medium grain	1 lb	454 g	2⅓ c	551 ml	193 g	6.8 oz
short grain	1 lb	454 g	2¼ c	532 ml	201 g	7.1 oz
Rice, all types						
cooked	1 lb	454 g	8 c	1.9 l	170 g	6.0 oz
precooked	8 oz	227 g			184 g	6.5 oz
prepared			2 c	473 ml	164 g	5.8 oz
parboiled	14 oz	397 g	2 c	473 ml	198 g	7.0 oz
Soy grits, stirred						
low-fat	1 lb	454 g	3 c	710 ml	150 g	5.3 oz
Teff	12 oz	340 g	1⅔ c	393 ml	198 g	7.0 oz
Wheat germ	12 oz	340 g	3 c	710 ml	113 g	4.0 oz

Buying Guide for Cereals and Flours (continued)

Food Item and Form	Market Unit		Approximate Volume or Pieces per Market Unit		Approximate Weight per Cup	
Flours						
Amaranth	1 lb	454 g	4^1/$_2$ c	1.1 l	102 g	3.6 oz
Corn	2 lb	908 g	8 c	1.9 l	116 g	4.1 oz
Gluten	2 lb	908 g	6^1/$_2$ c	1.5 l		
sifted					142 g	5.0 oz
Quinoa	24 oz	680 g	5^1/$_2$ c	1.3 l	122 g	4.3 oz
Rice	2 lb	908 g				
sifted			7 c	1.6 l	125 g	4.4 oz
stirred, spooned			5^3/$_4$ c	1.3 l	159 g	5.6 oz
Rye	2 lb	908 g				
light, sifted			10 c	2.4 l	88 g	3.1 oz
dark, sifted			7 c	1.6 l	128 g	4.5 oz
Soy	2 lb	908 g				
full-fat, sifted			15 c	3.5 l	60 g	2.1 oz
low fat			11 c	2.6 l	82 g	2.9 oz
Spelt	1 lb	454 g	4^1/$_4$ c	1.0 l	113 g	4.0 oz
Teff	12 oz	340 g	2^1/$_3$ c	551 ml	147 g	5.2 oz
Wheat						
all-purpose						
sifted	2 lb	908 g	8 c	1.9 l	116 g	4.1 oz
unsifted, spooned			7 c	1.6 l	125 g	4.4 oz
instant			7^1/$_4$ c	1.7 l	130 g	4.6 oz
bread, sifted	2 lb	908 g	8 c	1.9 l	113 g	4.0 oz
cake, sifted	2 lb	908 g	9^1/$_4$ c	2.2 l	96 g	3.4 oz
spooned			8^1/$_4$ c	2.0 l	110 g	3.9 oz
pastry, sifted	2 lb	908 g	9 c	2.1 l	99 g	3.5 oz
self-rising, sifted	2 lb	908 g	8 c	1.9 l	105 g	3.7 oz
whole wheat						
stirred	2 lb	908 g	6^2/$_3$ c	1.6 l	133 g	4.7 oz

Buying Guide for Cereals and Flours (continued)

Food Item and Form	Market Unit		Approximate Volume or Pieces per Market Unit		Approximate Weight per Cup	
Pastas						
Macaroni						
1″ pieces	1 lb	454 g	3³/₄ c	887 ml	123 g	4.3 oz
cooked			9 c	2.1 l	140 g	4.9 oz
Shell	1 lb	454 g	4-5 c	0.9-1.2 l	115 g	4.1 oz
cooked			9 c	2.1 l		
Noodles						
1″ pieces	1 lb	454 g	6-8 c	1.4-1.9 l	73 g	2.6 oz
cooked			8 c	1.9 l		
Spaghetti						
2″ pieces	1 lb	454 g	4-5 c	0.9-1.2 l	94 g	3.3 oz
cooked			9 c	2.1 l	159 g	5.6 oz
Starch						
Corn						
stirred	1 lb	454 g	3¹/₂ c	828 ml	128 g	4.5 oz
Potato						
stirred	1 lb	454 g	3¹/₄ c	769 ml	142 g	5.0 oz

Buying Guide for Fruits

Food Item and Form	Market Unit		Approximate Volume or Pieces per Market Unit		Approximate Weight per Cup	
Apples						
Fresh						
whole	1 lb	454 g	3 medium			
pared and sliced			2³/₄ c	651 ml	122 g	4.3 oz
sauce						
sweetened (not canned)			1³/₄ c	414 ml	252 g	8.9 oz
Frozen						
sliced,						
sweetened	20 oz	567 g	2¹/₂ c	592 ml	205 g	7.2 oz
Canned						
sliced	20 oz	567 g	2¹/₂ c	592 ml	231 g	7.5 oz
juice	46 fl oz	1.4 l	4³/₄ c	1.1 l	249 g	8.8 oz
sauce	1 lb	454 g	1³/₄ c	414 ml	259 g	9.1 oz
Dried	1 lb	454 g	4¹/₃ c	1.0 l	104 g	3.7 oz
cooked			8 c	1.9 l	244 g	8.6 oz

Buying Guide for Fruits (continued)

Food Item and Form	Market Unit		Approximate Volume or Pieces per Market Unit		Approximate Weight per Cup	
Apricots						
Fresh						
whole	1 lb	454 g	8 to 12		115 g	4.1 oz
sliced or halved			2½ c	592 ml	156 g	5.5 oz
Canned						
whole (medium)	1 lb	454 g	8 to 12		225 g	7.9 oz
halved (medium)	1 lb	454 g	12 to 20 halves		217 g	7.7 oz
Dried	11 oz	312 g	2¼ c	532 ml	150 g	5.3 oz
cooked, fruit						
and liquid			4⅓ c	1.0 l	285 g	10.0 oz
Avocado						
Fresh	1 lb	454 g				
sliced, diced,						
wedges			2½ c	592 ml	142 g	5.0 oz
Bananas						
Fresh						
whole	1 lb	454 g	3 to 4			
sliced			2 c	473 ml	142 g	5.0 oz
mashed			1⅓ c	315 ml	232 g	8.2 oz
Dried	1 lb	454 g	4½ c	1.1 l	100 g	3.5 oz
Blueberries						
Fresh	1 lb	454 g	2 c	473 ml	146 g	5.2 oz
Frozen	10 oz	284 g	1½ c	355 ml	161 g	5.7 oz
Canned	14 oz	397 g	1½ c	355 ml	170 g	6.0 oz
Cherries						
Fresh, red, pitted	1 lb	454 g	2⅓ c	551 ml	154 g	5.4 oz
Frozen, red, tart, pitted	20 oz	567 g	2 c	473 ml	242 g	8.5 oz
Canned, red, tart, pitted	1 lb	454 g	1½ c	355 ml	177 g	6.2 oz
Sweet, unpitted	1 lb	454 g	1¾ c	414 ml	177 g	6.2 oz
Cranberries						
Fresh						
uncooked	1 lb	454 g	4 c	946 ml	151 g	5.3 oz
sauce			4 c	946 ml	215 g	7.6 oz
Canned						
sauce	1 lb	454 g	1⅔ c	393 ml	278 g	9.8 oz
juice	1 qt	946 ml	4 c	946 ml	250 g	8.8 oz

Buying Guide for Fruits (continued)

Food Item and Form	Market Unit		Approximate Volume or Pieces per Market Unit		Approximate Weight per Cup	
Currants						
Dried	1 lb	454 g	3¼ c	769 ml	140 g	4.9 oz
Dates						
Dried, whole	1 lb	454 g	60 dates			
Pitted, cut	1 lb	454 g	2½ c	592 ml	178 g	6.3 oz
Figs						
Fresh	1 lb	454 g	12 medium			
Canned	1 lb	454 g	12 to 15 figs		230 g	8.1 oz
Dried						
whole	1 lb	454 g	44 figs			
cut fine			2⅔ c	551 ml	168 g	5.9 oz
Fruit Juice						
Frozen	6 fl oz	177 ml	¾ c	177 ml		
Canned	46 fl oz	1.4 l	5¾ c	1.4 l	247 g	8.7 oz
Fruits						
Frozen, mixed	12 oz	340 g	1⅓ c	315 ml		
Canned, cocktail						
or salad	17 oz	482 g	2 c	473 ml	229 g	8.1 oz
Grapefruit						
Fresh	1 lb	454 g	1 medium			
sections			1 c	237 ml	194 g	6.8 oz
Frozen sections	13.5 oz	383 g	1½ c	355 ml	219 g	7.7 oz
Canned sections	16 oz	454 g	2 c	473 ml	241 g	8.5 oz
Grapes						
Fresh						
seeded	1 lb	454 g	2 c	473 ml	184 g	6.5 oz
seedless	1 lb	454 g	2½ c	592 ml	169 g	6.0 oz
Kiwi						
Fresh	1 lb	454 g	3 medium			
Lemons						
Fresh	3 lb	1.4 kg	1 dozen			
Juice			2 c	473 ml	247 g	8.7 oz
Frozen juice	6 fl oz	177 ml	¾ c	177 ml	283 g	10.0 oz
Canned juice	8 fl oz	237 ml	1 c	237 ml	245 g	8.6 oz

Buying Guide for Fruits (continued)

Food Item and Form	Market Unit		Approximate Volume or Pieces per Market Unit		Approximate Weight per Cup	
Melons						
Frozen balls	12 oz	340 g	1½ c	355 ml	231 g	8.2 oz
Oranges						
Fresh	6 lb	2.7 kg	1 dozen			
diced or						
sectioned			12 c	2.8 l	214 g	7.5 oz
juice			4 c	946 ml	247 g	8.7 oz
Frozen juice						
reconstituted	6 fl oz	177 ml	3 c	710 ml	268 g	9.5 oz
Canned juice	46 fl oz	1.4 l	5¾ c	1.4 l	247 g	8.7 oz
Canned mandarin						
fruit and juice	11 oz	312 g	1¼ c	296 ml	250 g	8.8 oz
Peaches						
Fresh	1 lb	454 g	4 medium			
sliced			2 c	473 ml	177 g	6.2 oz
Frozen slices and juice	10 oz	284 g	1⅛ c	266 ml	251 g	8.8 oz
Canned						
halves	1 lb	454 g	6 to 10 halves		224 g	7.9 oz
slices	1 lb	454 g	2 c	473 ml	218 g	7.7 oz
Dried	1 lb	454 g	3 c	710 ml	160 g	5.6 oz
cooked	1 lb	454 g	6 c	1.4 l	244 g	8.6 oz
Pears						
Fresh	1 lb	454 g	4 medium			
sliced			2⅛ c	503 ml	213 g	7.5 oz
Canned halves	1 lb	454 g	6 to 10 halves		227 g	8.0 oz
Pineapple						
Fresh	2 lb	908 g	1 medium			
cubed			3 c	710 ml	146 g	5.2 oz
Frozen chunks	13.5 oz	383 g	1½ c	355 ml	204 g	7.2 oz
Canned						
chunks, tidbits	29 oz	822 g	3¾ c	887 ml	198 g	7.0 oz
crushed	29 oz	822 g	3¾ c	887 ml	260 g	9.2 oz
sliced	20 oz	567 g	10 slices		208 g	7.3 oz
juice	46 fl oz	1.4 l	5¾ c	1.4 l		

Buying Guide for Fruits (continued)

Food Item and Form	Market Unit		Approximate Volume or Pieces per Market Unit		Approximate Weight per Cup	
Plums						
Fresh	1 lb	454 g	8 to 20 plums			
halved			2 c	473 ml	185 g	6.5 oz
Canned, whole	1 lb	454 g	10 to 14 plums		223 g	7.9 oz
Prunes						
Canned	1 lb	454 g	10 to 14 prunes		196 g	6.9 oz
Dried						
whole	1 lb	454 g	2½ c	592 ml	176 g	6.2 oz
cooked			4-4½ c		229 g	8.1 oz
				946-1065 ml		
Dried						
pitted	1 lb	454 g	2¼ c	532 ml	162 g	5.7 oz
cooked			4-4½ c		210g	7.4 oz
				946-1065 ml		
Raisins						
Seeded						
whole	1 lb	454 g	3¼ c	769 ml	142 g	5.0 oz
chopped			2½ c	592 ml	182 g	6.4 oz
Seedless						
whole	1 lb	454 g	2¾ c	651 ml	146 g	5.2 oz
cooked			2¾ c	651 ml	183 g	6.5 oz
chopped			2 c	473 ml	189 g	6.7 oz
Rhubarb						
Fresh	1 lb	454 g	4 to 8 pieces			
cut					122 g	4.3 oz
cooked			2 c	473 ml	242 g	8.5 oz
Frozen, sliced	12 oz	340 g	1½ c	355 ml	168 g	5.9 oz
Strawberries						
Fresh						
whole	1.5 lb	680 g	4 c	946 ml	144 g	5.1 oz
sliced			4 c	946 ml	148 g	5.2 oz
Frozen						
whole	1 lb	454 g	1⅓ c	315 ml	204 g	7.2 oz
sliced or halved	10 oz	284 g	1 c	237 ml	235 g	8.3 oz

*Weight per cup is that of food alone without liquid, unless otherwise noted.

Buying Guide for Vegetables

Food Item and Form	Market Unit		Approximate Volume or Pieces per Market Unit		Approximate Weight per Cup (food alone, without liquid)	
Asparagus, spears						
Fresh	1 lb	454 g	16 to 20			
cooked			2 c	473 ml	181 g	6.4 oz
canned	15 oz	425 g	12 to 18		195 g	6.9 oz
Frozen spears,						
cuts, and tips	10 oz	284 g	2 c	473 ml	181 g	6.4 oz
Artichokes/Heart						
Bottoms			5 buttons			
canned	14 oz	397 g	1 c	237 ml	212 g	7.5 oz
Bamboo Shoots						
Canned	16 oz	454 g	1 c	237 ml	12 g	4.3 oz
Bean Sprouts						
Alfalfa, Mung	8 oz	227 g	3 c	710 ml	75 g	2.0 oz
Beans—Green						
Fresh	1 lb	454 g	3 c	710 ml	114 g	4.0 oz
cooked			2 1/2 c	592 ml	125 g	4.4 oz
Frozen	9 oz	255 g	1 1/2 c	355 ml	161 g	5.7 oz
Canned	16 oz	454 g	1 3/4 c	414 ml	135 g	4.8 oz
Beans—Kidney, Pinto, Red, Pink, Black						
dried	1 lb	454 g	2 1/4 c	532 ml	187 g	6.5 oz
canned	15-16 oz	425-454 g	2 c	473 ml	187 g	6.5 oz
dried, cooked	1 lb	454 g	5-6 c	1.2-1.4 l	185 g	6.5 oz
Beans—Lima						
shelled						
Fresh	1 lb	454 g	2 c	473 ml	155 g	5.5 oz
cooked			1 2/3-2 c		166 g	5.9 oz
				393-473 ml		
Frozen	10 oz	284 g	1 3/4 c	414 ml	173 g	6.1 oz
Canned	16 oz	454 g	2 c	473 ml	170 g	6.0 oz
Dried	1 lb	454 g	2 1/2 c	592 ml	180 g	6.3 oz
cooked			5 1/2 c	1.3 l	186 g	6.6 oz

Buying Guide for Vegetables (continued)

Food Item and Form	Market Unit		Approximate Volume or Pieces per Market Unit		ApproximateWeight per Cup (food alone, without liquid)	
Beans—Navy and Small White						
Dried	1 lb	454 g	2¹/₃ c	551 ml	190 g	6.7 oz
cooked	1 lb	454 g	5¹/₂ c	1.3 l	191 g	6.7 oz
Beans—Soybeans						
Dried	1 lb	454 g	2 c	473 ml	210 g	7.4 oz
Beets, without tops						
Fresh	1 lb	454 g	2 c	473 ml	145 g	5.1 oz
cooked			2 c	473 ml	180 g	6.3 oz
Canned	16 oz	454 g	2 c	473 ml	167 g	5.9 oz
Beet Greens	16 oz	454 g	2 c	473 ml	227 g	8 oz
Broccoli						
Fresh						
cooked	1 lb	454 g	2 c	473 ml	164 g	5.8 oz
Spears, chopped						
frozen	10 oz	284 g	1¹/₂ c	355 ml	188 g	6.6 oz
Brussels sprouts						
Fresh	1 lb	454 g	4 c	946 ml	102 g	3.6 oz
cooked			2¹/₂ c	592 ml	180 g	6.4 oz
Frozen	10 oz	284 g	18-24 sprouts			
	16 oz	454 g				
	20 oz	567 g				
Cabbage						
Fresh	1 lb	454 g				
shredded			3¹/₂-4¹/₂ c	0.8-1.1 l	80 g	2.8 oz
cooked			2 c	473 ml	146 g	5.2 oz
Cabbage, Chinese	1 lb	454 g	2 c	473 ml	146 g	5.2 oz

Buying Guide for Vegetables (continued)

Food Item and Form	Market Unit		Approximate Volume or Pieces per Market Unit		Approximate Weight per Cup (food alone, without liquid)	
Carrots, without tops						
Fresh	1 lb	454 g	3 c	710 ml	130 g	4.6 oz
shredded			2¹/₂ c	592 ml	112 g	4.0 oz
diced					137 g	4.8 oz
cooked			2-2¹/₂ c		160 g	5.6 oz
			473-592 ml			
Frozen	1 lb	454 g				
cooked			2¹/₂ c	592 ml	165 g	5.8 oz
Canned	16 oz	454 g	2 c	473 ml	159 g	5.6 oz
Cauliflower						
Fresh	1 lb	454 g	1¹/₂ c	355 ml	104 g	3.7 oz
cooked			1¹/₂ c	355 ml	125 g	4.4 oz
Frozen	10 oz	284 g	2 c	473 ml	152 g	5.4 oz
cooked			1¹/₂ c	355 ml	179 g	6.3 oz
Celery						
Fresh	1 lb	454 g	2 bunches		121 g	4.3 oz
Stalk	1 stalk		1 c	237 ml	150 g	5.3 oz
Cooked			2-2¹/₂ c		153 g	5.4 oz
			473-592 ml			
Corn						
Fresh ears	1 dozen					
cooked			5-6 c	1.2-1.4 l	165 g	5.8 oz
Frozen	10 oz	284 g	1³/₄ c	414 ml	135 g	4.8 oz
cut kernels	1 ear		¹/₂ c	118 ml	82 g	2.9 oz
cooked			1¹/₂-2 c		182 g	6.4 oz
			355-473 ml			
Canned						
cream style	16 oz	454 g	2 c	473 ml	249 g	8.8 oz
whole kernel	16 oz	454 g	2 c	473 ml	170 g	6.0 oz
Eggplant, fresh	1 lb	454 g				
diced			2¹/₂ c	592 ml	99 g	3.5 oz
cooked			2¹/₂ c	592 ml	213 g	7.5 oz
Greens, fresh	1 lb	454 g			77 g	2.7 oz
cooked			3 c	710 ml	190 g	6.7 oz
frozen	10 oz	284 g	1¹/₂-2 c		187 g	6.6 oz
			355-473 ml			
	16 oz	454 g				

Buying Guide for Vegetables (continued)

Food Item and Form	Market Unit		Approximate Volume or Pieces per Market Unit		Approximate Weight per Cup (food alone, without liquid)	
Juice						
Tomato	46 oz	1.4 kg	5³/₄ c	1.4 l	265 g	9.3 oz
Mixed vegetable	46 oz	1.4 kg	5³/₄ c	1.4 l	265 g	9.3 oz
Kolhrabi	1 lb	454 g				
Fresh, raw			1¹/₂ c slices	355 ml	140 g	4.9 oz
Lentils, dried	1 lb	454 g	2¹/₄ c	532 ml	191 g	6.7 oz
cooked			5 c	1.2 l	202 g	7.1 oz
Lettuce						
Head	1 lb	454 g	6¹/₄ c	1.5 l		
Leaf	1 lb	454 g	6¹/₄ c	1.5 l		
Romaine	1 lb	454 g	6 c	1.4 l		
Endive	1 lb	454 g	4¹/₄ c	1.0 l		
Mixed vegetables						
Frozen	10 oz	284 g	2 c	473 ml	182 g	6.4 oz
	16 oz	454 g				
	32 oz	908 g				
Canned	16 oz	454 g	2 c	473 ml	179 g	6.3 oz
Mushrooms						
Fresh, sliced	1 lb	454 g	2-3 c	473-710 ml	68 g	2.4 oz
Canned	4 oz	113 g	²/₃ c	156 ml	161 g	5.7 oz
Okra						
Fresh, cooked	1 lb	454 g	2¹/₄ c	532 ml	177 g	6.2 oz
Frozen	10 oz	284 g	1¹/₄ c	296 ml	209 g	7.4 oz
Canned	14¹/₂ oz	411 g	1³/₄ c	414 ml	171 g	6.0 oz
Onions						
Fresh	1 lb	454 g	3 large			
chopped			2-2¹/₂ c		135 g	4.8 oz
				473-592 ml		
Cooked					197 g	6.9 oz
Frozen, chopped	12 oz	340 g	3 c	710 ml		
Canned	16-17 oz		2 c	473 ml		
		454-482 g				
Dried					64 g	2.3 oz

Buying Guide for Vegetables (continued)

Food Item and Form	Market Unit		Approximate Volume or Pieces per Market Unit		Approximate Weight per Cup (food alone, without liquid)	
Parsnips						
Fresh	1 lb	454 g	4 medium			
cooked			2 c	473 ml	211 g	7.4 oz
Peas,green						
Fresh, in pod	1 lb	454 g				
shelled			1 c	237 ml	138 g	4.9 oz
cooked			1 c	237 ml	163 g	5.7 oz
Frozen	10 oz	284 g	2 c	473 ml	156 g	5.5 oz
cooked			2 c	473 ml	167 g	5.9 oz
Canned	17 oz	482 g	2 c	473 ml	168 g	5.9 oz
Dried, split	1 lb	454 g	2¼ c	532 ml	200 g	7.1 oz
cooked			5 c	1.2 l	194 g	6.8 oz
Peas, black-eyed						
Fresh	1 lb	454 g			144 g	5.1 oz
cooked			2⅓ c	551 ml	162 g	5.7 oz
Frozen, cooked	16 oz	454 g	1½ c	355 ml	171 g	6.0 oz
Canned	15 oz	425 g	2 c	473 ml	205 g	7.2 oz
Dried, split	1 lb	454 g			200 g	7.1 oz
cooked					248 g	8.7 oz
Peppers—Sweet, Bell, Green, Red, Yellow						
Fresh	1 lb	454 g	3 medium			
strips			3½ c	828 ml	114 g	4.0 oz
diced			3 c	710 ml	126 g	4.4 oz
Cooked strips			2½-3 c	592-710 ml	138 g	4.8 oz
Poi	1 lb	454 g	1 c	237 ml	240 g	8.4 oz

Buying Guide for Vegetables (continued)

Food Item and Form	Market Unit		Approximate Volume or Pieces per Market Unit		Approximate Weight per Cup (food alone, without liquid)	
Potatoes, white						
Fresh	1 lb	454 g	3 medium		164 g	5.8 oz
cooked, diced						
or sliced			2¼ c	532 ml	163 g	5.7 oz
mashed			1¾ c	414 ml	207 g	7.3 oz
Frozen, french						
fried or puffs	20 oz	567 g	4½ c	1.1 l	113 g	4.0 oz
	32 oz	908 g	8 c	1.9 l	113 g	4.0 oz
Canned, whole	15-16 oz	425-454 g	8-12 c	1.9-2.8 l	179 g	6.3 oz
Dried flakes	6-7 oz	170-198 g	4½ c	1.1 l	36 g	1.3 oz
reconstituted			10¾ c	2.5 l	212 g	7.5 oz
Dried granules	1 lb	454 g	2¼ c	532 ml	201 g	7.1 oz
reconstituted			10½ c	2.5 l	212 g	7.5 oz
Hash browns						
frozen	10 oz	284 g	4 c	946 ml	114 g	4.0 oz
	32 oz	908 g	8 c	1.9 l	114 g	4.0 oz
Pumpkin						
Fresh						
cooked, mashed	1 lb	454 g	1 c	237 ml	247 g	8.7 oz
Canned	16 oz	454 g	2 c	473 ml	244 g	8.6 oz
	29 oz	822 g	3 c	710 ml	244 g	8.6 oz
Radishes, sliced	6 oz	170 g	1¼ c	296 ml		
Rutabaga						
Fresh						
cubed	1 lb	454 g	2½ c	592 ml	139 g	4.9 oz
cooked			2 c	473 ml	163 g	5.7 oz
Saukerkraut						
Canned	16 oz	454 g	3 c	710 ml	188 g	6.6 oz
	29 oz	822 g	5½ c	1.3 l	188 g	6.6 oz
Spinach						
Fresh	1 lb	454 g	3 c	710 ml	54 g	1.9 oz
cooked			1 c	237 ml	200 g	7.1 oz
Frozen	10 oz	284 g	1½ c	355 ml	190 g	6.7 oz
Canned	15 oz	425 g	2 c	473 ml	221 g	7.8 oz

Buying Guide for Vegetables (continued)

Food Item and Form	Market Unit		Approximate Volume or Pieces per Market Unit		Approximate Weight per Cup (food alone, without liquid)	
Squash, winter						
Fresh	1 lb	454 g				
cooked, mashed			1 c	237 ml	244 g	8.6 oz
Frozen	12 oz	340 g	1½ c	355 ml	242 g	8.5 oz
Canned, drained						
Solids	16 oz	454 g	1¾ -2 c			
				414-473 ml		
Squash, summer						
Fresh	1 lb	454 g			136 g	4.8 oz
cooked, mashed			1⅔ c	393 ml	238 g	8.4 oz
Frozen	10 oz	284 g	1½ c	355 ml	211 g	7.4 oz
Canned	1 lb	454 g				
Sweet potatoes/Yams						
Fresh	1 lb	454 g	3 medium			
cooked, sliced					232 g	8.2 oz
Frozen	12 oz	340 g	3-4		200 g	7.1 oz
Canned	16 oz	454 g	1¾-2 c		220 g	7.8 oz
				414-473 ml		
	29 oz	822 g	3½ c	828 ml	220 g	7.8 oz
Dried, flakes	1 lb	454 g			115 g	4.1 oz
reconstituted					255 g	9.0 oz
Tomatoes						
Fresh	1 lb	454 g	3-4 small		162 g	5.7 oz
cooked			1½ c	355 ml		
Canned, whole	16 oz	454 g	2 c	473 ml	238 g	8.4 oz
	28 oz	794 g	3½ c	828 ml	238 g	8.4 oz
Sauce	8 oz	227 g	1 c	237 ml	258 g	9.1 oz
	16 oz	454 g	2 c	473 ml	258 g	9.1 oz
Turnips						
Fresh	1 lb	454 g	3 medium		134 g	4.7 oz
cooked			2 c	473 ml	196 g	6.9 oz

Buying Guide for Spices and Herbs

Food Item and Form	Market Unit		Approximate Volume per Market Price		Approximate Weight per Cup	
Allspice	1.1 oz	31 g	1.5 T	22 ml	96 g	3.4 oz
Anise	0.7 oz	20 g	3 T	45 ml	108 g	3.8 oz
Basil					71 g	2.5 oz
Bay Laurel	0.1 oz	3 g			28 g	1.0 oz
Caraway					108 g	3.8 oz
Cardamom					94 g	3.3 oz
Cayenne Pepper ground	2 oz	57 g	½ c	118 ml	85 g	3.0 oz
Celery Seed	1.6 oz	45 g	7 T	105 ml	105 g	3.7 oz
Chervil					28 g	1.0 oz
Chili Powder	1.5 oz	42 g	5 T	75 ml	119 g	4.2 oz
Cinnamon, ground	1.6 oz	45 g	6 T	90 ml	108 g	3.8 oz
Clove, ground	1.4 oz	40 g	5 T	75 ml	105 g	3.7 oz
Coriander Seed, ground	1.7 oz	48 g	9 T	135 ml	79 g	2.8 oz
Cumin Seed	1.4 oz	40 g	6 T	90 ml	96 g	3.4 oz
Dill Seed	1.6 oz	45 g	7 T	105 ml	105 g	3.7 oz
Fennel	1.8 oz	51 g	9 T	135 ml	94 g	3.3 oz
Ginger					85 g	3.0 oz
Mace, ground					85 g	3.0 oz
Marjoram, ground	0.9 oz	26 g			28 g	1.0 oz
Mustard Seed	2.4 oz	68 g	6 T	90 ml	179 g	6.3 oz
Nutmeg, ground	1.3 oz	37 g	6 T	90 ml	110 g	3.9 oz
Oregano, ground	1 oz	28 g	6 T	90 ml	71 g	2.5 oz
Paprika	1.4 oz	40 g	5 T	75 ml	110 g	3.9 oz
Parsley	0.2 oz	6 g	5 T	75 ml	20 g	0.7 oz
Pepper, black	4 oz	113 g	1 c	237 ml	102 g	3.6 oz
Poppy Seed	1 oz	28 g	3 T	45 ml	142 g	5.0 oz
Rosemary	0.2 oz	6 g	2 T	30 ml	54 g	1.9 oz
Saffron	0.05 oz	1 g	2 T	30 ml	34 g	1.2 oz
Sage, ground	0.4 oz	11 g	6 T	90 ml	31 g	1.1 oz
Savory, ground					71 g	2.5 oz
Sesame Seed	2.1 oz	60 g	7 T	105 ml	128 g	4.5 oz
Tarragon					76 g	2.7 oz
Thyme, ground	0.9 oz	26 g	6 T	90 ml	68 g	2.4 oz
Turmeric, ground	0.7 oz	20 g	3 T	45 ml	108 g	3.8 oz

Buying Guide for Sweetening Agents

Food Item and Form	Market Unit		Approximate Volume per Market Price		Approximate Weight per Cup	
Sugar						
Brown, packed						
light	1 lb	454 g	2¼ c	532 ml	201 g	7.1 oz
dark	1 lb	454 g	2¼ c	532 ml	201 g	7.1 oz
granulated	1 lb	454 g	3 c	710 ml	153 g	5.4 oz
Cane or beet						
granulated	5 lb	2.3 kg	11¼ c	2.7 l	201 g	7.1 oz
ultrafine	2 lb	908 g	4⅔ c	1.1 l	196 g	6.9 oz
confectioner's						
unsifted	1 lb	454 g	3-4 c	710-946 ml	122 g	4.3 oz
sifted			4½ c	1.1 l	96 g	3.4 oz
Corn Syrup						
Light & dark	16 fl oz	473 ml	2 c	473 ml	329 g	11.6 oz
Honey	1 lb	454 g	1⅓ c	315 ml	332 g	11.7 oz
Maple Syrup	12 fl oz	355 ml	1½ c	355 ml	312 g	11.0 oz
Molasses						
Cane	12 fl oz	355 ml	1½ c	355 ml	309 g	10.9 oz
Sorghum	1 lb	454 g	1⅓ c	315 ml	329 g	11.6 oz
Sugar						
Fructose	5.3 oz	150 g	50 packets		194 g	6.8 oz
	10.0 oz	284 g	2 c	473 ml		
High Intensity Sweeteners						
Saccharin*	3.5 oz	99 g	1 g packets		142 g	5.0 oz
	2.8 oz	80 g	1 g packets		29 g	1.0 oz
Aspartame	7 oz	198 g	1 g packets		159 g	5.6 0z
	2 oz	56.7 g	2⅜ c	557 ml		
Acesulfame						
Potassium	1.75 oz	50 g	1 g packets		144 g	5.1 oz

*Brands of saccharin sweeteners vary in weight per cup.

Buying Guide for Leavening Agents

Food Item and Form	Market Unit		Approximate Volume per Market Price		Approximate Weight in Grams	
					per tsp	per Tbsp
Baking powder						
SAS-Phosphate	14 oz	397 g	2½ c	592 ml	3.8 g	11.4 g
Low sodium or						
sodium-free	8 oz	227 g	1⅓ c	315 ml	3.5 g	10.5 g
Baking soda	1 lb	454 g	2⅓ c	551 ml	4.0 g	12.2 g
Cream of tartar	1.75 oz	50 g	5¼ Tbsp	79 ml	3.1 g	9.4 g
Yeast						
Active dry &						
quick	0.25 oz	7 g	1 Tbsp	15 ml	2.3 g	7.0 g
Compressed	0.60 oz	17 g	4 tsp	20 ml	4.2 g	12.8 g

Buying Guide for Miscellaneous Foods[1]

Food Item and Form	Market Unit[1]		Approximate Volume per Market Price		Approximate Weight per Cup	
Bouillon						
cube	3.25 oz	92 g	25 cubes			
powder	6.25 oz	177 g			177 g	6.25 oz
Bread						
Sliced	1 lb	454 g	12-16 slices			
crumbs						
soft			10 c	2.4 l	45 g	1.6 oz
dry	10 oz	283 g	2¾ c	651 ml	102 g	3.6 oz
Croutons, cubed	6 oz	170 g			51 g	1.8 oz
Catsup, tomato	14 oz	397 g	1½ c	355 ml	272 g	9.6 oz
Chocolate						
Bitter/semisweet chips	8 oz	227 g	1 c	237 ml	224 g	7.9 oz
	12 oz	340 g	2 c	473 ml	196 g	6.9 oz
Prepared drink			30 c	7.1 l		

[1] The market unit may reflect only one arbitrarily selected market unit. The market place now permits considerable variety.

Buying Guide for Miscellaneous Foods[1] (continued)

Food Item and Form	Market Unit		Approximate Volume per Market Price		Approximate Weight per Cup	
Cocoa	8 oz	227 g	2 c	473 ml	113 g	4.0 oz
Prepared drink			50 c	11.8 l		
Instant	8 oz	227 g	1²/₃ c	393 ml	139 g	4.9 oz
prepared drink			28 c	6.6 l		
Coconut						
long thread	1 lb	454 g	5²/₃ c	1.3 l	79 g	2.8 oz
canned, moist	1 lb	454 g	5 c	1.2 l	85 g	3.0 oz
Coffee	1 lb	454 g	5 c	1.2 l	85 g	3.0 oz
Brewed			40-50 c	9.5-11.8 l		
Instant	2 oz	57 g	1¹/₄-1¹/₂ c	296-355 ml	40 g	1.4 oz
brewed			60 c	14 l		
Freeze-dried	4 oz	113 g	2 c	473 ml	54 g	1.9 oz
beverage	10 oz	284 g	40 tsp	200 ml		
Crackers						
Graham	1 lb	454 g	66			
crumbs			4¹/₃ c	1.0 l	85 g	3.0 oz
Soda	1 lb	454 g	82			
crumbs			7 c	1.6 l		
fine	10 oz	284 g	4 c	946 ml	71 g	2.5 oz
Saltines	1 lb	454 g	130-140			
Gelatin						
Granulated						
unflavored	1 oz	28 g	¹/₄ c	59 ml	150 g	5.3 oz
flavored	3 oz	85 g	7 Tbsp	105 ml	179 g	6.3 oz
prepared			2 c	473 ml	269 g	9.5 oz
Infant foods						
Strained and						
junior (chopped)	3.25-3.5 oz	92-99 g	6 Tbsp	90 ml		
	4.25-4.75 oz	120-135 g	9 Tbsp	135 ml		
	7.5-8 oz	213-284 g	15 Tbsp	225 ml		
Juice	4 fl oz	118 ml	¹/₂ c	118 ml		

[1] The market unit may reflect only one arbitrarily selected market unit. The market place now permits considerable variety.

Buying Guide for Miscellaneous Foods[1] (continued)

Food Item and Form	Market Unit		Approximate Volume per Market Price		Approximate Weight per Cup	
Mayonnaise	1 pt	473 ml			244 g	8.6 oz
Nuts, shelled						
Almonds						
blanched	1 lb	454 g	3 c	710 ml	153 g	5.4 oz
Filberts						
whole	1 lb	454 g	3½ c	828 ml	133 g	4.7 oz
Peanuts	1 lb	454 g	3 c	710 ml	144 g	5.1 oz
Pecans						
halved	1 lb	454 g	4 c	946 ml	108 g	3.8 oz
chopped	1 lb	454 g	3½-4 c		119 g	4.2 oz
				828-946 ml		
Pistachio	1 lb	454 g	3¼-4 c		125 g	4.4 oz
				769-946 ml		
Walnuts						
Persian, English						
halved	1 lb	454 g	3½ c	828 ml	99 g	3.5 oz
chopped	1 lb	454 g	3½ c	828 ml	119 g	4.2 oz
Paste, Almond	8 oz	227 g				
Peanut butter	18 oz	510 g	2 c	473 ml	252 g	8.9 oz
Pudding						
instant mix	6 oz	170 g	13 Tbsp	195 ml	210 g	7.4 oz
Salad dressing						
French	1 pt	473 ml	2 c	473 ml	249 g	8.8 oz
Salt						
free-running	1 lb	454 g	1½ c	355 ml	289 g	10.2 oz
Soups						
Frozen						
condensed	10-10.5 oz	284-298 g	1-1½ c	237-355 ml		
Ready-to-serve	15 oz	425 g	1½-2 c	355-473 ml		
Canned						
condensed	10.5-11.5 oz	298-326 g	1¼ c	296 ml		
prepared			2½ c	591 ml		
ready-to-serve	8 oz	227 g	1 c	237 ml	227 g	8.0 oz
Dried	2.75 oz	78 g				
reconstituted			3 c	710 ml	232 g	8.2 oz

[1] The market unit may reflect only one arbitrarily selected market unit. The market place now permits considerable variety.

Buying Guide for Miscellaneous Foods[1] (continued)

Food Item and Form	Market Unit		Approximate Volume per Market Price		Approximate Weight per Cup	
Tapioca						
Quick cooking	8 oz	227 g	1½ c	355 ml	153 g	5.4 oz
Tea, leaves	1 lb	454 g	6⅓ c	1.5 l	71 g	2.5 oz
Brewed			300 c	71 l		
Instant	1.5 oz	42 g	1¼ c	296 ml	71 g	2.5 oz
brewed			64 c	15 l		
Herbal	2 oz	57 g	24 bags			
Water					237 g	8.4 oz
Whipped toppings						
dry mix	1.3 oz	37 g	6 Tbsp	90 ml	108 g	3.8 oz

[1] The market unit may reflect only one arbitrarily selected market unit. The market place now permits considerable variety.

Volume and Weight of Commercial Size Containers

Can Size		Net Weight	Products Packaged	Approximate Measure	
6 oz	6 oz	170 g	Frozen juices	0.75 c	177 ml
8 oz	8 oz	227 g	concentrates, individual serving juices, vegetables and fruits, tomato paste and sauce	1 c	237 ml
No. 1 Picnic	10½ oz	298 g	Condensed soups, some fruits, vegetables, meat and fish products	1.25 c	296 ml
No. 2 Vacuum	12 oz	340 g	Vegetables	1.50 c	355 ml
	14½ oz	411 g	Evaporated milk		
	15 oz	425 g	Sweetened condensed milk		

Volume and Weight of Commercial Size Containers (continued)

Can Size	Net Weight		Products Packaged	Approximate Measure	
No. 300	15½ oz	439 g	Pork and beans, spaghetti, macaroni, chili, date and nut bread, cranberry sauce and blueberries	1.75 c	414 ml
No. 303	15 oz	425 g	Vegetables, fruits	2 c	473 ml
No. 2	20 oz (1 lb 4 oz)	567 g	Juices, vegetables, fruits	2.5 c	592 ml
No. 2½	29 oz (1 lb 13 oz)	822 g	Fruits, vegetables	3.5 c	828 ml
No. 3	46 oz (1 qt 14 fl oz)	1.4 l	Juices whole chicken	5.75 c	1.4 l
No. 10	6 lb 9 oz	3 kg	Institutional or restaurant size for many canned products	3 qt	2.8 l

Food Preservation

DRYING

Some state Cooperative Extension services offer bulletins with instructions for drying foods and for constructing dehydrators. There are a few accurate commercial books available on how to dry foods. Consider the following facts before drying is begun:

- Most fruits, some vegetables, meats, poultry, fish, herbs, nuts, and seeds dry well, but they must be fresh and of high quality to produce a quality product.

- Pretreatment of foods prior to dehydrating is essential. Methods of pre-treating include: ascorbic acid treatment, salt (sodium chloride) solution, or blanching. Blanching can be either in water or steam. Steam blanching is preferred over boiling water for some vegetables since more water soluble nutrients are retained. Steam blanching requires 50 percent more time than water blanching. Blanching will inactivate enzymes and retard vitamin C loss. Sulfur treatment is not recommended because some individuals are sensitive to it. However, it is useful in preventing discoloration in some fruits, such as apricots. Pretreatment is required for most fruits and vegetables.

- Drying racks may be homemade or commercial, but they should not include such harmful metals as galvanized or cadmium-coated products or copper.

- Drying temperatures must be high enough to prevent foodborne illness, but not so high as to cause deterioration of the product. Generally, temperatures between 120°F and 140°F (50°C–60°C) are used. For best results, use 140°F (60°C).

- Testing for dryness should be done when the food is at room temperature, as hot food will respond differently.

- Conditioning the food in a large glass jar for one week after drying will ensure evenness and thoroughness in drying.

- Dried foods are drastically reduced in volume, often one-quarter of the original volume. As a result, they are often consumed in large amounts since they seem small and light-weight, without attention given to their high caloric content and nutrient density.

- Dried foods store well in glass jars in a cool, dry, dark place for maximum shelf life. Containers should be of appropriate size for the amount of food to keep air space to a minimum. Plastic bags sometimes do not protect from insect damage.

- When rehydrating dried foods, any thin liquid that combines well with the food can be used. Examples of liquids are: water, milk, fruit juices, bouillon or ginger ale. For best results with fruit, pour just enough boiling liquid over them to cover. For vegetables, cover and soak in a cold liquid until nearly restored to their original texture. Cover meat with boiling water or broth. Let sit for 30 to 60 minutes, followed by 15 to 20 minutes of simmering before using.

- Regardless of the equipment or techniques used in drying food, vitamin C loss will be great.

- Microwave ovens are not recommended for drying. The food tends to burn or dry excessively in one spot while being moist in others, and the oven may be damaged.

FOOD FOR DRYING

Most fruits dry well. Those which do not dry well include: avocados and olives because of high fat content; berries (other than strawberries), guavas, and pomegranates because of high seed content; citrus fruits because of juice content and soft pulp; crabapples, quince, and muskmelon because of objectionable flavors when dried. Berries need to be sieved to remove seeds prior drying to make fruit to leather.

- Select firm, ripe fruits. Bruised fruits can often be used in fruit leather after bruises are cut out, but are not suitable for halves or slices.

- Drying times may vary due to humidity, load, type of food and efficiency of the dehydrator. The usual range is 3–5 hours for small fruits, 4–7 hours for other fruits except pineapple (8–12 hours).

- Many vegetables are difficult to dry or to rehydrate satisfactorily. Carrots, corn, mushrooms, onions, parsnips, peas and peppers dry well. These are inexpensive and available fresh year-round, but drying home-grown vegetables is practical and convenient.

- Select absolutely fresh, mature vegetables of high quality. They should be washed well just before drying.

- Drying times will vary from 1–2 hours for herbs, 2–4 hours for other vegetables. The thinner the piece, the shorter the drying time.

- Meat, poultry, or fish should be lean because fat becomes rancid during storage. Meat must be very fresh. Eggs should not be dried in a home dehydrator as quality will be poor.

- Collect herbs for drying as follows: Leaves should be collected just before the plant flowers; flowers should be collected just after the blossoms have opened; seeds should be collected when mature and no longer green.

- Select freshly picked herbs, gathered just after the dew is off. Herbs dry well at room temperature if household dust is not a problem.

- Nuts and seeds are foods naturally preserved by drying. They are usually used in dried form for making desserts and salads or eaten as snacks.

CANNING

Organisms that cause food spoilage—molds, yeast, and bacteria—are always present in the air, water, and soil. Enzymes that may cause undesirable changes in flavor, color, and texture are present in raw fruits and vegetables. When food is canned, it must be processed at a high enough temperature and for a period that is long enough to destroy any spoilage organisms. This heating process also stops the action of enzymes. The method of canning is determined by the pH or acidity of the food.

PRESSURE CANNING

This processing method, utilizing a pressure canner (not a pressure saucepan or steamer), is used for low-acid foods. At sea level, all vegetables (except pickled vegetables), all meat, poultry, and seafood are processed at 10 pounds of pressure with a weighted gauge and 11 pounds with a dial gauge (70 kPa, 240°F or 116°C). Above 1,000 feet in elevation the pressure must be increased to offset the decreasing atmospheric pressure. The higher the elevation, the less atmospheric pressure, hence the lower the boiling point of water. Lower boiling temperatures are less effective in killing bacteria. Increased pressure raises the boiling temperature. *Clostridium botilinum* spores can survive boiling water canner temperatures, but are killed by the high temperature reached in a pressure canner when processing follows the recommended time and pressure. The pressure canner dial gauge should be tested for accuracy each canning season. County extension offices can tell you where to have your gauge tested. Dial gauges should be replaced if they read inaccurately.

There are two methods of packing vegetables into jars. Check for specific vegetables in Table 13.1.

Raw Pack—Pack cold, raw vegetables tightly into canning jars or cans and cover with boiling water, wipe the jar rim, and seal with the lid. Place jars in the pressure canner.

Hot Pack—Prepare vegetables by boiling in water a short time (varies with specific vegetable). Loosely fill canning jars with hot vegetable and then add liquid, leaving 1-inch headspace. Wipe the jar rim and adjust lid.

Processing—Follow the manufacturer's directions for the canner you are using.

TABLE 13.1

Timetable for Home Canning Vegetables in Pressure Canner[1,2]

Vegetable	Packing Method	Pint Jar	Quart Jar
Asparagus, 1″ (2.5 cm) pieces	raw or hot	30 min	40 min
Beans, green, 1″ (2.5 cm) pieces	raw or hot	20 min	25 min
Beans, lima, fresh	raw or hot	40 min	50 min
Beets, small or sliced	hot	30 min	35 min
Carrots, sliced or diced	raw or hot	25 min	30 min
Corn, cream style	hot	85 min	not recommended
Corn, whole kernel	raw or hot	55 min	85 min
Mushrooms, small, whole or large, halved or quartered	hot	45 min	not recommended
Okra, cut or whole	hot	25 min	40 min
Peas, green	raw or hot	40 min	40 min
Peppers	hot	35 min	not recommended
Potatoes, small whole	hot	35 min	40 min
Pumpkin, 1″ (2.5 cm) cubes	hot	55 min	90 min
Pumpkin, strained	not recommended		
Spinach	hot	70 min	90 min
Squash, winter, 1″ (2.5 cm) cubes	hot	55 min	90 min
Squash, winter, strained	not recommended		
Sweet potatoes, cooked	hot	65 min	90 min
Tomato juice	hot	15 min	15 min
Tomatoes, whole or halved	raw or hot	25 min	25 min

[1] at 10 pounds in a weighted gauge canner; 11 pounds in a dial gauge canner (70 kPa, 240°F or 116°C).

[2] Pressure must be adjusted for altitude by increasing the pressure 1 pound (6.9 kPa) for each 2,000 feet (609 m) above sea level for the dial gauge canner and to 15 pounds (103.4 kPa) for all altitudes above 1,000 feet (304 m) for the weighted-gauge pressure canner.

Here are a few pointers on the use of most pressure canners:

■ Preheat two to three inches of water to 180°F (82°C) in the bottom of the canner.

■ Set filled jars on rack in pressure canner so that steam can flow around each jar.

■ Fasten lid securely so that no steam can escape except through vent (petcock or weighted-gauge opening).

■ Watch until steam pours steadily from vent as directed. Close petcock or place weighted gauge on vent.

■ Let pressure rise to appropriate pounds required for processing. (See Table 13.1) The moment this pressure is reached, start counting processing time. Keep pressure constant by regulating heat under the canner. Do not lower pressure by opening petcock. Keep drafts from blowing on canner.

■ When processing time is up, remove canner from heat immediately. Let canner stand at room temperature until pressure is zero. Never try to rush the cooling. When pressure registers zero, wait two minutes, then slowly open petcock. Unfasten lid and tilt the far side up so steam escapes away from you. Remove jars from canner.

When canning meat, poultry and seafood, purchase fresh, high quality products and process within 24 hours. When processing homegrown meat or poultry, processing should occur within 24–48 hours after slaughter. The hot pack method of pressure canning is recommended.

- Place large, cut-to-measure pieces of boned, defatted meat in a large shallow pan. Addjust enough water to keep meat from sticking. Cover and cook slowly on top of the range or in a 350°F (180°C) oven until the meat is medium done or it reaches an internal temperature of 155°F (68°C), turning it occasionally so it precooks evenly.

- Pack hot meat loosely in jar, leaving one inch (2.5 cm) headspace. Add 1 teaspoon (5 ml) salt per quart (1 l) if desired. Add boiling meat juice (extended with boiling water if necessary), leaving one inch (2.5 cm) of headspace. Wipe jar rims thoroughly; adjust lids according to manufacturer's instructions.

- Pressure-process at 90 minutes at 10 pounds pressure with a weighted gauge; 11 pounds with a dial gauge (70 kPa, 240°F or 116°C). Be sure to adjust for altitude.

BOILING WATER CANNER METHOD

This processing method is used for all acid foods: fruits, tomatoes with added acid, jams, jellies, and pickled vegetables. The canner must be at least four inches (10cm) higher than the tops of the jars. It needs a lid and wire rack. A pressure canner may be used, with the petcock left wide open to prevent pressure buildup. See Table 13.2 for correct processing time. As is also true for pressure canning, never use a jar unless it is a canning jar.

Pack food either raw or hot:

Raw Pack—Put raw fruits into container and cover with boiling hot syrup, juice, or water. Press tomatoes down in the containers so they are covered with their own juice. See directions listed under hot pack for acidifying tomatoes.

Hot Pack—Heat fruits in syrup, water, or steam, or in extracted juice before packing. Juicy fruits and tomatoes may be preheated without added liquid and packed in the juice of cooked fruit. To acidify tomatoes add two tablespoons (30 ml) of bottled lemon juice or 1/2 teaspoon (3 ml) of citric acid per quart (l) of tomatoes.

Sugar syrup concentrations used in canning fruits are as follows:

- Fill jars, leaving 1/2 inch (1.2 cm) headspace. Put on cap, screw band firmly. Put filled and capped glass jars into half-filled water bath canner of preheated water, 140°F (60°C) for raw pack and 180°F (82°C) for hot-packed foods.

- Add boiling water, if needed, to bring water an inch or two (2.5–5 cm) over tops of jars. Do not pour boiling water directly on glass jars. Put lid on canner.

- When water in canner comes to a rolling boil, start to count processing time. See Table 13.2. Boil gently and steadily for time recommended for the food being canned. Add boiling water during processing, if needed, to keep jars covered. Immediately remove jars from the canner when processing time is complete.

- Cool overnight before testing for seal and removing screws bands. Wash jars, dry, label, and store without bands.

Sugar Type	Sugar %	Water		Sugar		Yield	
				c	ml	c	l
Very light	10	6 1/2	1.6	3/4	175	6 3/4	1.6
Light	20	5 3/4	1.4	1 1/2	375	6 1/2	1.5
Medium	30	5 1/4	1.3	2 1/4	550	6	1.4
Heavy	40	5	1.2	3 1/4	800	6 1/2	1.5
Very heavy	50	4 1/4	1.0	4 1/4	1050	6 1/2	1.5

TABLE 13.2

Boiling Water Canner Processing

Food Item	Style of Pack	Jar Size	Time in Minutes			
			0–1000[1] (0–304 m)	1001–3000[1] (303–913 m)	3001–6000[1] (914–1826 m)	Above 6000[1] (1826 m +)
Apple slices	hot	pt	20	25	30	35
		qt	20	25	30	35
Applesauce	hot	pt	15	20	20	25
		qt	20	25	30	35
Apricots, peaches, nectarines	hot	pt	20	25	30	35
		qt	25	30	35	40
	raw	pt	25	30	35	40
		qt	30	35	40	45
Berries	hot	pt	15	20	20	25
		qt	15	20	20	25
	raw	pt	15	20	20	25
		qt	20	25	30	35
Cherries	hot	pt	15	20	20	25
		qt	20	25	30	35
	raw	pt	25	30	35	40
		qt	25	30	35	40
Fruit puree except figs and tomatoes	hot	pt	15	20	20	25
		qt	15	20	20	25
Grapefruit, orange sections	raw	pt	10	15	15	20
		qt	10	15	15	20
Pears	hot	pt	20	25	30	35
		qt	25	30	35	40
Plums	hot/raw	pt	20	25	30	35
		qt	25	30	35	40
Rhubarb	hot	pt	15	20	20	25
		qt	15	20	20	25
Tomato						
juice	hot	pt	35	40	45	50
		qt	45	50	55	60
whole/halved water pack	hot/raw	pt	40	45	50	55
		qt	45	50	55	60
whole/halved juice pack	hot/raw	pt	85	90	95	100
		qt	85	90	95	100
no added liquid	raw	pt	85	90	95	100
		qt	85	90	95	100
Jams and jellies	hot	1/2 pt/pt	5	10	10	15

TABLE 13.2 (CONTINUED)
Boiling Water Canner Processing

Food Item	Style of Pack	Jar Size	Time in Minutes			
			0–1000[1] (0–304 m)	1001–3000[1] (303–913 m)	3001–6000[1] (914–1826 m)	Above 6000[1] (1826 m +)
Spreads	hot	1/2 pt/pt	15	20	20	25
			20	30	30	35
Dill pickles	raw	pt	10	15	15	20
		qt	15	20	20	25
Bread & butter pickles	hot	pt	10	15	15	20
		qt	10	15	15	20
14-day sweet pickles	raw	pt	5	10	10	15
		qt	10	15	15	20
Sauerkraut	hot	pt	10	15	15	20
		qt	15	20	20	25
	raw	pt	20	25	30	35
		qt	25	30	35	40

See Table 13.3 for estimation of yield and jars needed for canning fruits and vegetables.

PICKLES AND RELISHES

For all pickle and relish products, processing in the boiling water canner is recommended. See Table 13.2 for recommended times. **DO NOT USE OPEN KETTLE CANNING.**

- Use vinegar with five to six percent acidity. Canning and pickling salt should be used for all brines.

- Processing times for relishes depend on the combination of vegetables. Obtain a USDA recipe to insure safety.

- For brined or fermented pickles, use a USDA recipe and a fermenting time of two to four weeks.

- For fresh pack pickles, cure them in brine for several hours or overnight.

- For relishes prepared from fruits and vegetables, chop, season, and then cook. Hot pack.

- For fruit pickles, prepare them from whole fruits and simmer in a spicy, sweet-sour syrup. Hot pack.

CAUTION: Developing your own recipes and using combinations of foods are not recommended in home canning. If you choose to do this, freeze the product. Only recipes based on recommendations of USDA should be used. Changing ingredients can result in different density and acidity resulting in an under processed and unsafe food.

JELLIES AND JAMS

Jelly, jam, conserve, marmalade, and preserves are all fruit products that add zest to meals. These methods provide a way to use inferior fruit, the largest or smallest fruits and berries and those that are irregularly shaped, that are not suitable for canning or freezing. Basically, these products are similar; all of them are preserved by means of sugar and usually gelled to some extent. Pectin, sugar, and acid levels of fruit vary, making it impossible to develop formulas that will always give the same results. In products made with pectin, use 1/4 to 1/2 cup (50 to 125 ml) more fruit or juice to make a softer product. Use ¹/₄ to ¹/₂ cup (50 to 125 ml) less fruit or juice to make a firmer product. In products made without added pectin, a shorter cooking time gives a softer product. Lengthen the cooking time for a firmer product.

Jelly is made from fruit juice; the product is clear and firm enough to hold its shape when turned out of the container.

Jam is made from crushed or ground fruit. It tends to hold its shape but generally is less firm than jelly.

Freezer jam is made with added pectin and is not processed. The taste and color are more like fresh fruit.

Conserves are jams made from a mixture of fruits, usually including citrus fruits; often raisins and nuts are added.

Marmalade is a tender jelly with small pieces of fruit distributed evenly throughout. A marmalade commonly contains citrus fruit.

Preserves are whole fruits or large pieces of fruit in a thick syrup, often slightly gelled.

Fruit butter is a smooth, less sweet preserve made from fruit pulp run through a sieve and cooked slowly.

Some fruits have enough natural pectin to make high-quality products. These fruits are: citrus, some apples, grapes, blackberries, boysenberries, cranberries and some plums. Other fruits require added pectin for making jellies that are firm enough to hold their shape. These fruits are: peaches, pears, apricots, elderberries, strawberries and raspberries. All fruits have less pectin when they are fully ripe than when they are underripe.

Commercial fruit pectins made from apples or citrus fruits are on the market in both liquid and powdered form. Either form is satisfactory when used in a recipe developed especially for that form. Commercial pectins can lose their gelling capabilities if stored too long or cooked too long when preparing the product. Unused pectin should be stored in a cool dry place.

The advantages of using pectin are that fully ripe fruit can be used, cooking time is short, and the yield is greater.

The acid content varies in different fruits and is higher in underripe fruits. With fruits low in acid, lemon juice or citric acid is commonly added to make gelled products. Commercial fruit pectins contain some acid. Lemon juice is usually added to increase acid content of fruit. If desired, ¹/₈ teaspoon (0.6 ml) of citric acid can be substituted for each tablespoon (15 ml) of lemon juice.

When using cane or beet sugar, the ratio of sugar to fruit is ³/₄:1 for most fruits. It is best to add sugar before cooking. Sugar helps in gel formation, serves as a preserving agent, and contributes to the flavor. It also has a firming effect on fruit, a property that is useful in the making of preserves. To use corn syrup or honey in jelly making, use a tested recipe to ensure success.

Gelling tests for making jelly without pectin are:

Spoon or sheet test: dip a cool metal spoon in boiling jelly, then carry it out of the steam and pour from the spoon. Jelly is done when syrup forms two drops that flow together and sheet off the edge of the spoon.

Freezer test: remove jelly from heat. Pour a small amount of boiling jelly on a cold plate and place in freezer. When it is cooled, see if it is gelled.

To make jelly with pectin using the short boil method, mix powdered pectin with unheated fruit juice, bring mixture to a boil, add sugar and boil for one minute at a full boil. If using liquid pectin, add it to the boiling juice and sugar mixture and boil for one minute at a full rolling boil. Follow directions for the specific product.

Filling and sealing containers when jelly is done:

For jelly, fill to ¹/₄ inch (0.6 cm) from top and seal. Process according to Table 13.2.

For sweet spreads, let stand 5 minutes, stirring occasionally to keep fruit from floating. Place in jars and

TABLE 13.3

Estimation of Yield and Jars Needed for Canning Fruits and Vegetables[1]

Raw Produce	Measures	Approximate Number Quart (950 ml) Jars Needed	Approximate Pounds Needed for 1 Quart Jar
Fruits			
Apples	1 bu. (48 lb)	16–20	2$\frac{1}{2}$ to 3
Applesauce	1 bu. (48 lb)	15–18	2$\frac{1}{2}$ to 3$\frac{1}{2}$
Apricots	1 lug (22 lb)	7–11	2 to 2$\frac{1}{2}$
Berries	24 quart crate	12–18	1$\frac{1}{2}$ to 3
Cherries	1 bu. (56 lb)	22–33 (unpitted)	2 to 2$\frac{1}{2}$
	1 lug (22 lb)	9–11 (unpitted)	2 to 2$\frac{1}{2}$
Peaches	1 bu. (48 lb)	18–24	2 to 3
	1 lug (22 lb)	8–12	2 to 3
Pears	1 bu. (50 lb)	20–25	2 to 3
	1 box (35 lb)	14–17	2 to 3
Plums	1 bu. (56 lb)	24–30	1$\frac{1}{2}$ to 2$\frac{1}{2}$
	1 lug (24 lb)	12	1$\frac{1}{2}$ to 2 $\frac{1}{2}$
Tornatoes	1 bu. (53 lb)	15–20	2$\frac{1}{2}$ to 3 $\frac{1}{2}$
	1 lug (30 lb)	10	2$\frac{1}{2}$ to 3$\frac{1}{2}$
Tomatoes for juice	1 bu. (53 lb)	12–16	3 to 3$\frac{1}{2}$
Vegetables			
Beans, lima (in pods)	1 bu. (32 lb)	6–10	3 to 5
Beans, green or wax	1 bu. (30 lb)	12–20	1$\frac{1}{2}$ to 2$\frac{1}{2}$
Beets (without tops)	1 bu. (52 lb)	15–24	2 to 3$\frac{1}{2}$
Carrots (without tops)	1 bu. (50 lb)	16–25	2 to 3
Corn, sweet		6–10	3 to 6
(in husks)	1 bu. (35 lb)	(whole kernel)	
Okra	1 bu. (26 lb)	16–18	1$\frac{1}{2}$
Peas (in pods)	1 bu. (30 lb)	5–10	3 to 6
Spinach and other greens	1 bu. (18 lb)	3–8	2 to 6
Squash, summer	1 bu. (40 lb)	10–20	2 to 4
Sweet potatoes	1 bu. (50 lb)	18–25	2 to 3

NOTE: The actual number of jars needed in canning depends on the size and condition of the produce and the manner of preparing and packing into jars. The standard weight of a bushel, lug or box is not the same in all states.

[1]*Ball Blue Book*, Edition 32, Ball Corporation, Muncie, Indiana, 1992.

seal. Then process in boiling water canner according to Table 13.2.

Reduced-sugar fruit spreads are made with several different gelling or thickening agents, each of which gives somewhat different characteristics to the product. Reduced or no sugar may be used. Sweet fruits, apple juice, and/or a liquid low or non-caloric sweetener may be added for flavor. Preservation is dependent on boiling water canner processing, refrigerator storage, or frozen storage. Directions given for the commercial thickening ingredient being used should be followed. If unflavored gelatin powder is used as the gelling agent, the products must be refrigerated and used within 4 weeks.

FREEZING

Freezing preserves many foods with little change in flavor, color, texture, and nutritional value. Freezing, one of the simplest methods of home food preservation, does not sterilize the product; it simply retards the growth of harmful bacteria and molds. Therefore, good sanitation procedures are necessary for freezing. The wrapping or container for the food that is being frozen is very important. Boxes, jars, and bags especially made for freezing are available. A freezer container or wrap should:

- be impermeable to oxygen
- be odorless
- possess high wet strength
- be greaseproof
- be nonadherent to the frozen product
- be puncture resistant
- be easy to mark
- not crack at low temperature.

Plastic freezer containers or other sealed rigid containers are preferred for liquid packs. Canning jars designed for freezing used with home canning lids are excellent for freezing liquids. Jars are easily resealed after removing a portion of their contents. Allow enough headspace for expansion of liquid during freezing to prevent breakage. Be sure to follow manufacturer's directions.

Flexible bags or wrappings of moisture vapor resistant polyethylene, cellophane, nylon, or plastic are suitable for dry-pack vegetables, fruits, meats, poultry, and fish. Bags may also be used for liquid packs, but breakage and leakage should be expected.

Laminated freezer papers made of various combinations of paper, foil or plastic are suitable for dry-pack meats, fish, and poultry. Metal foils are easily torn or punctured.

BASIC STEPS IN FREEZING

- Use only quality fresh products.
- Work under very sanitary conditions.
- Organize everything that is needed to save time and energy.
- Use appropriate packing materials.
- Follow directions for the product you are freezing: blanch or scald vegetables; blanch 1 minute longer at 5000 feet (1524 m) and above; cool quickly use ascorbic acid for light colored fruits to prevent discoloration use appropriate pack for fruits keep meats cold while preparing for freezing.
- Package, removing all air.
- Label with date and name of product.
- Promptly place in freezer in single layers.
- Maintain 0°F (–18°C) in freezer.
- Keep an inventory.
- Use all products within recommended storage period for best quality.

How To Pack Fruits

Fruits should be slightly riper than for canning. Fruits can be packed in syrup or dry sugar. The intended use of the fruit will determine the type of pack used. Forty percent syrup is recommended for most fruits. See Table 13.4 for quantities needed.

- Syrup pack: dissolve sugar in water until solution is clear. Use just enough cool syrup to cover the fruit in the container. Allow 1/2 inch (1.2 cm) headspace.

Type of Syrup %	Sugar c	Sugar l	Water c	Water l	Amount of Syrup c	Amount of Syrup l
30	2	0.5	4	1	5	1.2
40	3	0.8	4	1	$5^{1}/_{2}$	1.4
50	$4^{3}/_{4}$	1.2	4	1	$6^{1}/_{2}$	1.6
60	7	1.8	4	1	$7^{3}/_{4}$	1.9

- Sugar pack: works well for soft fruits such as strawberries. Add sugar to clean, prepared fruits. Let stand 10 to 15 minutes. Stir to coat each piece. Pack fruit tightly in storage containers.

- Dry pack: pack fruit in containers without sugar or syrup. No headspace is necessary.

- Individual quick frozen (IQF): this pack is good for small whole fruit. The fruit is frozen individually without sugar on a shallow metal tray or cookie sheet, then packed in a container for storage.

Ascorbic acid should be used with fruits that tend to darken quickly (peaches, apricots, figs, sweet cherries, pears, and apples). Follow manufacturer's directions as to method and amount.

Freezing fruit allows jelly and jam making anytime of the year.

Refer to Table 13.4 for information on percent recovery.

VEGETABLES

Speed from garden to freezer is the important factor in vegetable freezing. Blanching destroys enzymes that will cause deterioration of quality and vitamin C loss during storage. Blanching also cleanses the surface of microorganisms and softens the product which makes it easier to package. Vegetables are lowered into boiling water or in steam for a specific amount of time and then quickly cooled before packing to freeze. (See Table 13.5 for blanching times).

Vegetables for freezing can be packed in two ways.

TABLE 13.4

**Weight Conversions:
Farm to Processed Frozen Fruits[1]**

Frozen Fruits	% Recovery	Frozen Weight from Farm Weight[2]
Apples	60	0.60
Apricots	78	0.91
Berries		
Blackberries	95	0.95
Blueberries	97	0.97
Boysenberries	88	0.88
Gooseberries	97	0.97
Loganberries	88	0.88
Raspberries	95	0.95
Strawberries	93	1.12
Cherries		
sour	75	0.90
sweet	85	0.85
Grapes	85	0.85
Peaches	67	0.80
Pineapples	50	0.625
Prunes	85	0.85

[1] Economic, Statistics, and Cooperative Services, U.S. Department of Agriculture. From *The Almanac of Canning, Freezing, Preserving and Allied Industries.* Edward E. Judge & Sons, Inc., Westminster, MD.

[2] Frozen weight is weight of frozen fruits plus sugar content. Fruit to sugar ratio is used in this computation: apricots 6 to 1; strawberries, sour cherries, peaches 5 to 1; pineapples 4 to 1; other fruits 0.

TABLE 13.5

Timetable for Blanching Vegetables Prior to Freezing

Vegetable	Minutes Heated in Boiling Water [1,2]	Vegetable	Minutes Heated in Boiling Water [1,2]
Asparagus		medium ears—1¼ to 1½	
small stalks	1½	inches (3-3.8 cm) in diameter	6
medium stalks	2	large ears—1½ to 2 inches	
large stalks	3	(3-5 cm) in diameter	8
Beans, lima		extra large ears—over 2 inches	
small beans or pods	1	(5 cm) in diameter	10
medium beans or pods	2	whole kernel	5 to 6
large beans or pods	3	**Greens, all types**	2
Beans, wax or green	3	**Kohlrabi**	
Beets		½ inch (1.2 cm) cubes	2
small	25 to 30	whole, small to medium	3
medium	45 to 50	**Okra**	
Broccoli		small pods	3
florets 1½ inches (3.8 cm)		large pods	5
in diameter	3	**Parsnips,** ½ inch (1.2 cm) cubes	
large	4	or slices	5
Brussels sprouts		**Peas**	
small heads	3	black-eyed	
medium	4	small	1
large	5	large	2
Cabbage		green	2
coarse shreds or thin wedges	1½	**Pumpkin**	until soft
wedges	3	**Rutabagas,** ½ inch (1.2 cm) cubes	3
Carrots		**Squash**	
whole small	5	summer, 1½ inch (1.2 cm) slices	3
diced, sliced, strips	3	winter	until soft
Cauliflower		**Sweet potatoes**	until soft
medium	3	**Tomatoes**	
large	4	whole	until soft
Corn, sweet		juice, simmer tomatoes	5
small ears—1¼ inches (3 cm)		**Turnips,** ½ inch (1.2 cm) cubes	3
or less in diameter	5		

[1] For each pound of prepared vegetable, use at least 1 gallon of boiling water. After heating the specified time, cool promptly in cold water and drain.

[2] Adjust time for altitude so under-blanching does not occur.

Dry Pack—After vegetables have been blanched, cooled, and drained, place them in meal-size, airtight, moisture-proof containers. After packaging, remove all air possible from the container before it is sealed. Allow headspace for expansion of the vegetables during freezing.

Individual quick frozen (IQF)—After vegetable pieces have been blanched, cooled, and drained, spread them in a single layer on shallow trays or pans. Place trays in the freezer for four to six hours, then quickly package frozen pieces in a suitable resealable container. Unblanched diced onions and peppers may be IQF packed.

MEAT

Freezing is the best method of meat storage. Starting with a wholesome, high-quality product will ensure the best results. All meat should be wrapped in moisture-vapor-proof material that is impermeable to oxygen.

THAWING

With the exception of spinach and corn-on-the-cob, vegetables may be taken directly from the freezer, unpacked, and steamed or boiled. Spinach and corn-on-the-cob are better if first thawed in a microwave oven or four hours in the refrigerator.

Meats, fish, shellfish, and vegetables must not be thawed at room temperature. Many bacteria on fresh food will still be present when food is thawed. The nutrients in the drip that forms upon thawing support rapid bacterial growth. Bacteria capable of causing illness are common on vegetables, meats, poultry, and seafood. Improper thawing of these products may lead to illness. Thawing sealed packages of foods rapidly in cold water is an acceptable practice, but the cold water should be changed every 5 to 10 minutes. Proper thawing will help ensure safe food. Leftover frozen foods should be held below 40°F or above 140°F (4°C–60°C) to avoid possible foodborne illnesses.

FOODS NOT SUITABLE FOR FREEZING

Some foods not suitable for freezing include raw cabbage, cake batters, raw celery, cream pie fillings, custards, cooked egg whites, lettuce, mayonnaise, potato salad, salad dressing, fresh tomatoes, and watermelon.

IF FREEZER STOPS

If power fails or the freezer stops operating normally, try to determine how long it will be before the freezer will be back in operation.

A fully-loaded freezer usually will stay cold enough to keep foods frozen for two days if the cabinet is not opened. In a cabinet with less than half a load, food may not stay frozen more than one day. Some partially thawed foods can be refrozen. Generally, refreezing should not pose a health hazard if the food has not reached 40°F (4°C). If thawed food still contains a few ice crystals, it is safe to refreeze. Quality loss occurs with thawing and refreezing. However, it is least evident in red meats and most evident in fish and seafood.

Meat, Poultry and Vegetables—Refreeze if the temperature is 40°F (4°C) or below and if the color and odor are good. Discard packages that show signs of spoilage, such as off-colors or bad odors.

Fruits—Refreeze if they smell and taste good. Thawed fruits may also be used in cooking or making jellies, jams, and preserves.

Cooked foods—Refreeze if ice crystals are still present or the freezer is 40°F (4°C) or below. If the condition is poor or questionable, throw it out.

If normal operation cannot be resumed before the food will start to thaw, use dry ice. If dry ice is placed in the freezer soon after the power is off, 25 pounds (11 kg) should keep the temperature below freezing for two to three days in a 10-cubic-foot (0.3 m³) cabinet with half a load, and three to four days in a loaded cabinet. Place the dry ice on cardboard or small boards on top of packages and do not open cabinet again except to put in more dry ice or to remove it when normal operation is resumed.

Another suggestion is to move the food to a locker plant, using insulated boxes or thick layers of paper to prevent thawing while in transit. The locker plant can refreeze the packaged food more quickly, resulting in better quality.

Sources of Information

In today's world, web addresses and content of websites are constantly changing. Although the information in this section may be current at the time of publication, readers will need to be responsible for updating this information.

U. S. GOVERNMENT RESOURCES

Gateway to government food safety information:
www.FoodSafety.gov

U.S. Food and Drug Administration
Center for Food Safety and Applied Nutrition
200 C St., SW
Washington, DC 20204
vm.cfsan.fda.gov

Center for Complementary and Alternative Medicine
National Institutes of Health
9000 Rockville Pike
Bethesda, MD 20892
nccam.nih.gov
e-mail: info@nccam.nih.gov

Center for Nutrition Policy and Promotion
U.S. Department of Agriculture
1120 29th St., NW, North Lobby
Suite 200
Washington, DC 20036
202-418-2312
www.usda.gov/cnpp

Centers for Disease Control and Prevention (CDC)
1600 Clifton Rd., NE
Atlanta, GA 30333
800-311-3435
www.cdc.gov

Discussion of American Eating Habits
800-999-6779
www.ers.usda.gov/Publications/aib750

FDA Food Information and Seafood Hotline
888-SAFE-FOOD (723-3366)

FDA/HAACP
Site for Hazard Analysis Critical Control Point
vm.cfsan.fda.gov/~lrd/haccp.html

FDA Office of Biotechnology
200 C St., SW
Washington, DC 20204
888-INFO-FDA (devices, drugs)
also: USDA has a site on agricultural biotechnology:
www.aphis.usda.gov/subjects/biotechnology

Federal Trade Commission
(tips on fraudulent practice)
www.ftc.gov

Food and Drug Administration (FDA)
Consumer Information Office (HFE-88)
5600 Fishers Ln.
Rockville, MD 20857
888-INFO-FDA (888-463-6332)
(or contact your regional FDA office)
www.fda.gov

Food and Nutrition Information Center
National Agricultural Library
U.S. Department of Agriculture
Room 304
10301 Baltimore Ave.
Beltsville, MD 20705
301-504-5719
www.nal.usda.gov/fnic
Many food and nutrition-related links, including those to MyPyramid, Dietary Guidelines, and food composition.

Food Safety and Inspection Service
U.S. Department of Agriculture
14th St and Independence Ave., SW
Washington, DC 20250
202-720-2791
www.foodsafety.gov
Frequently updated; the gateway to government food safety information.

International Food Information Council (IFIC)
1100 Connecticut Ave., NW
Suite 430
Washington, DC 20036
202-296-6540
e-mail: foodinfo@ific.org
www.ific.org

Metric Program
National Institute of Standards and Technology
100 Bureau Dr.
Stop 2000
Gaithersburg, MD 20899-2000
301-975-3690
ts.nist.gov/ts/htdocs/200/202/mpo_home.htm

National Academy of Sciences
Food and Nutrition Board
2101 Constitution Ave., SW
Washington, DC 20418
202-334-1732
nationalacademies.org,snb@nas.edu

National Association of Nutrition and Aging Services
Programs/ National Meals on Wheels
1100 Vermont Ave., NW
Suite 1001
Washington, DC 20005
202-682-6899
http://nanasp.org

National Lead Information Center
8601 Georgia Ave.
Suite 503
Washington, DC 20910
800-424-LEAD (5323)
www.epa.gov/lead/pubs/nlic.htm

Safe Drinking Water Hotline
Environmental Protection Agency
800-426-4791

United States Department of Agriculture (USDA)
www.usda.gov

USDA – Agricultural Research Service
Beltsville Human Nutrition Research Service
Beltsville, MD 20705
www.barc.usda.gov
e-mail: bhnrc@bhnrc.arsusda.gov
www.ars.usda.gov/ba/bhnrc/ndl

USDA – Agricultural Research Service
Nutrient Data Laboratory
(contains the USDA Nutrient Database for Standard Reference, Release 12. Lists values for 100 grams, edible portion, instructions for downloading)
www.nal.usda.gov/fnic/foodcomp

USDA Cooperative State Research, Education, and Extension Service (CSREES); Cooperative Extension System (CES)
612-626-1111
www.cyfernet.org
e-mail: cyf@reeusda.gov

USDA – Food Guide Pyramid for Young Children
www.usda.gov/cnpp/KidsPyra/

USDA Meat and Poultry Hotline
Food Safety and Inspection Service
Department of Agriculture
Washington, DC 20250
888-674-6854
mphhotline.fsis@usda.gov

USDA School Meals
schoolmeals.nal.usda.gov
Lists recipes, menus, other information for school meals.

PROFESSIONAL ORGANIZATIONS

American Association of Cereal Chemists
3340 Pilot Knob Rd.
St. Paul, MN 55121-2097
651-454-7250
www.aaccnet.org
e-mail: aacc@scisoc.org

**American Association of Family
and Consumer Sciences**
Nutrition, Health and Food Management Division
400 N. Columbus St., Ste. 202
Alexandria, VA 22314
800-424-8080; 703-706-4600
www.aafcs.org
e-mail: staff@aafcs.org

American Dietetic Association
National Center for Nutrition and Dietetics
216 West Jackson Blvd.
Chicago, IL 60606-6995
www.eatright.org

Consumer Nutrition Hotline:
800-366-1655, to listen to recorded messages in English
or Spanish or to obtain a referral to an RD in your area

American Public Health Association
800 I St., NW
Washington, DC 20001-3710
202-777-2742
www.apha.org
e-mail: comments@apha.org

Herb Research Foundation
1007 Pearl St., Ste. 200
Boulder, CO 80302
303-449-2265
voice mail: 800-748-2617
www.herbs.org

Institute of Food Technologists
221 N. LaSalle St.
Suite 300
Chicago, IL 60601-1291
1-800-IFT-FOOD (438-3663); 312-782-8424
www.ift.org
e-mail: info@ift.org

International Association of Culinary Professionals
304 West Liberty St.
Suite 201
Louisville, KY 40202
502-581-9786
www.iacp.com
e-mail: iacpa@hqtrs.com

**National Association of College and
University Food Service**
1405 South Harrison Blvd.
Suite 305
Manly Miles Bldg.
Michigan State Univ.
East Lansing, MI 48824
517-332-2494
www.nacufs.org

School Nutrition Association
700 S. Washington St.
Suite 300
Alexandria, VA 22314
1-800-877-8822; 703-739-3900
www.schoolnutrition.org
e-mail: servicecenter@schoolnutrition.org

Society for Nutrition Education
1001 Connecticut Ave.
Suite 528
Washington, DC 20036
800-235-6690; 202-452-8534
www.sne.org
e-mail: info@sne.org

INDUSTRY-SPONSORED ORGANIZATIONS, TRADE ASSOCIATIONS

Food Marketing Institute (contains extensive list of
links to food, nutrition, and government websites)
655 15th St., NW
Suite 700
Washington, DC 20005
202-452-8444
www.fmi.org
e-mail: fmi@fmi.org

National Restaurant Association
1200 17th St., NW
Washington, DC 20036
800-424-5156; 202-331-5990
www.restaurant.org
e-mail: info@restaurant.org

Produce for a Better Health Foundation
5 A Day Program
5301 Limestone Rd.
Suite 101
Wilmington, DE 19808-1249
302-235-ADAY (2329)
www.5aday.com

AGRICULTURE COMMODITY BOARDS AND COMMISSIONS, PRODUCT PROMOTION ASSOCIATIONS

American Dry Bean Board
www.americanbean.org

Northarvest Bean Growers Association
50072 East Lake Seven Rd.
Frazee, MN 56544
218-334-6351
www.northarvestbean.org/html/consumer.cfm
e-mail: nhlbean@loretel.net

American Egg Board
1460 Renaissance Dr.
Suite 301
Park Ridge, IL 60068
847-296-7043
www.aeb.org

Egg Nutrition Center
1050 17th St., NW
Suite 560
Washington, DC 20036
202-833-8850
www.enc-online.org
e-mail: enc@enc-online.org

American Soybean Association
12125 Woodcrest Executive Dr.
Suite 100
St. Louis, MO 63141
314-576-1770
www.soygrowers.com

Cattlemen's Beef Board
5420 S. Quebec St.
Greenwood Village, CO 80111-1905
303-694-0305
www.beefitswhatsfordinner.com
e-mail: cattle@beef.org

Food Products Association
1350 I St., NW, Suite 300
Washington, DC 20005
202-639-5900
www.fpa-food.org

Glutamate Association
202-783-6135
www.msgfacts.com

Home Baking Association
10841 S. Crossroads Dr.
Suite 105
Parker, CO 80138
303-840-8787
www.homebaking.org
e-mail: HBAPatton@aol.com

International Bottled Water Association
1700 Diagonal Rd.
Suite 650
Alexandria, VA 22314
800- WATER11; 703-683-5213
www.bottledwater.org

International Dairy Foods Association
1250 H St., NW
Suite 900
Washington, DC 20005
202-737-4332
www.idfa.org

International Food Information Council
1100 Connecticut Ave., NW
Suite 430
Washington, DC 20036
202-296-6540
www.ific.org
e-mail: foodinfo@ific.org

International Food Information Service (IFIS)
Lane End House
Shinfield Rd.
Shinfield
Reading RG2 9BB
UK
Food Science and Technology Abstract providers
(commercial site with numerous links)
www.foodsciencecentral.com
e-mail: IFIS@IFIS.ORG

Mushroom Council
11875 Dublin Blvd.
Suite D262
Dublin, CA 94568
www.mushroomcouncil.org
e-mail: info@mushroomcouncil.org

National Chicken Council
1015 15th St., NW
Suite 930
Washington, DC 20005
202-296-2622
www.eatchicken.com

National Coffee Association
15 Maiden Lane
Suite 1405
New York, NY 10038
212-766-4007
www.ncausa.org

National Dairy Council/American Dairy Association
www.ilovecheese.com
also: www.nationaldairycouncil.org

Milk Processor Education Program
whymilk.com

National Fisheries Institute, Inc.
1901 N. Fort Myer Dr.
Suite 700
Arlington, VA 22209
703-524-8880
www.aboutseafood.com

National Food Service Management Institute
The University of Mississippi
6 Jeanette Phillips Drive
P.O. Drawer 188
University, MS 38677-0188
www.olemiss.edu

National Pasta Association
www.ilovepasta.org

National Pork Producers Council
P.O. Box 10383
Des Moines, IA 50306
515-223-2600
www.nppc.org

National Turkey Federation
1225 New York Ave., NW
Suite 400
Washington, DC 20005
202-898-0100
www.eatturkey.com
e-mail: info@turkeyfed.org

The Partnership for Food Safety Education
API/Fight Bac
4550 Forbes Blvd.
Lanham, MD 20706
301-731-6101
www.fightbac.org
e-mail: info@fightbac.org

Produce Marketing Association
1500 Casho Mill Rd.
P.O. Box 6036
Newark, DE 19714-6036
302-738-7100
www.pma.com
e-mail: webmaster@mail.pma.com

Salt Institute
700 N. Fairfax St.
Suite 600
Alexandria, VA 22314-2026
703-549-4648
www.saltinstitute.org

Snack Food Association
1711 King St.
Suite One
Alexandria, VA 22314
800-628-1334; 703-836-4500
www.sfa.org
e-mail: sfa@sfa.org

The Sugar Association, Inc.
1101 15th St., NW
Suite 600
Washington, DC 20005
202-785-1122
www.sugar.org

Tea Association of the U.S.A.
420 Lexington Ave.
New York, NY 10170
212-986-6998
www.teausa.org

United Fresh Fruit & Vegetable Association
727 N. Washington St.
Alexandria, VA 22314
703-836-3410
www.uffva.org

United Soybean Board
16640 Chesterfield Grove Rd.
Suite 130
Chesterfield, MO 63005-1429
800-989-USB1 (8721)
www.unitedsoybean.org

U.S.A. Rice Federation
4301 North Fairfax Dr.
Suite 305
Arlington, VA 22203
703-351-8161
www.usarice.com

U.S. Poultry and Egg Association
1530 Cooledge Rd.
Tucker, GA 30084-7303
770-493-9401
www.poultryegg.org
e-mail: webmaster@poultryegg.org

Washington State Apple Commission
www.bestapples.com

Wheat Foods Council
10841 South Crossroads Dr.
Suite 105
Parker, CO 80138
303-840-8787
www.wheatfoods.org

SOME ADDITIONAL WEBSITES

Body Mass Index (BMI) — will calculate BMI with your weight and height; also provides nutritional components of many foods, including fast foods: nutrio.com

The Cook's Thesaurus
www.foodsubs.com

Culinary Links (extensive list):
www.ehotelier.com/browse/culinary.php

Culinary and Wine Schools:
www.shawguides.com

Digital Chef:
www.digitalchef.com

Epicurious:
food.epicurious.com

Food and Wine:
http://www.globalgourmet.com

Food Network:
www.foodnetwork.com

Joy of Baking:
www.joyofbaking.com

Metric conversion:
www.convert-me.com

Metropolitan Life height and weight tables,
eating tips:
www.metlife.com

Restaurants and institutions:
www.foodservice411.com/rimag

Reynolds Kitchen:
www.reynoldskitchen.com

Rhodes Bread:
www.rhodesbread.com (800-876-7333)

Equipment

Alto-Shaam:
www.alto-shaam.com

Frymaster:
www.frymaster.com

Groen:
difc.difoodservice.com

Hobart:
www.hobartcorp.com

Market Forge:
www.mfii.com

Nasco:
www.nascofa.com

Oneida:
www.oneida.com

Presto:
www.presto-net.com

Industry Service Careers
Aramark:
www.aramark.com

Compass Group:
www.cgnad.com

Sodexho:
www.sodexhousa.com

Index

Won ton, defined, 63

Y

Yeast, 134
    compressed, 134
    dry yeast, 134
    quick rise, 134
    substitute for, 32
Yeast breads
    baking temperatures for, 30
    basic recipe, 35
    high altitude and, 13
Yeast rolls, baking temperatures
    for, 30

Yield of home canned products, 174
Yield of home frozen products, 176
Yogurt, 68
    buying of, 145
    frozen, 68
Yorkshire pudding, defined, 63